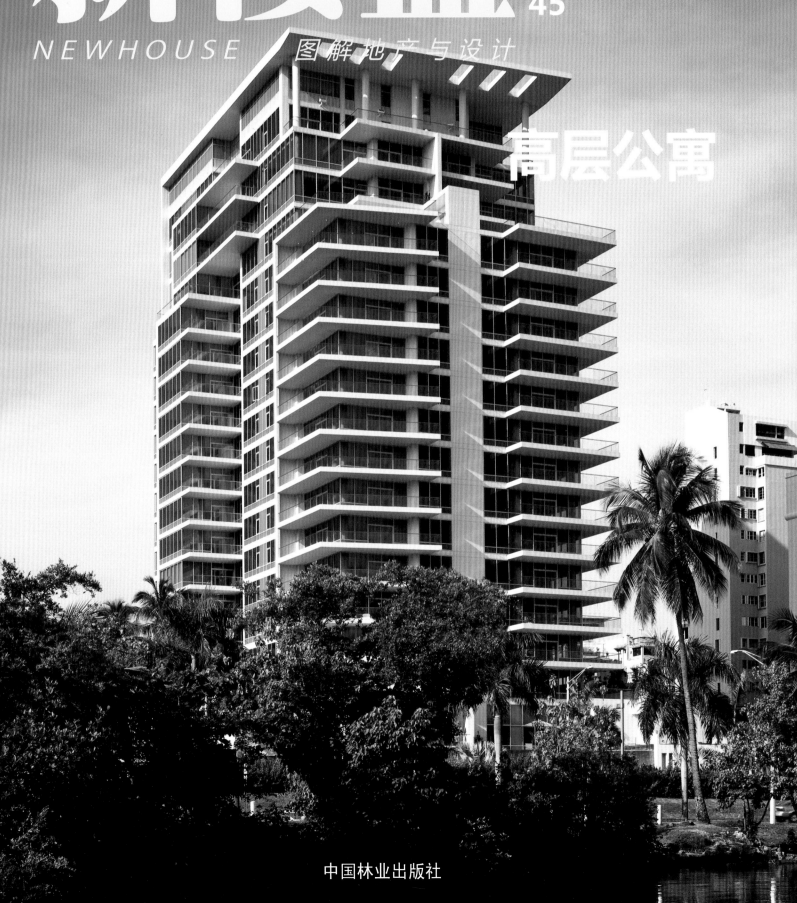

新楼盘
NEWHOUSE 图解地产与设计

45

高层公寓

中国林业出版社

天萌（中国）建筑设计机构
TEAMER ARCHITECTURAL DESIGN AND CONSULTANTS CO.,LTD

认识我们　About Us

天萌（中国）——建筑设计梦工场。四位始萌于国际设计团队的担纲者都有着二十余年的从事一线建筑设计的经验，在国内直接经历并参与了中国房地产业近二十年非凡的发展，见证了中国的城市化进程，对建筑与自然融合及文化传承有着深刻的认识及充分的实践积累。

天萌是"Team"，是团队、集约、高效的标志。

天萌是"力量"，"句者毕出"，"萌者尽达"。

天萌（中国）建筑设计机构注册于国内广州，公司拥有各专业设计精英百余人，十多年的骨干团队仅近几年已完成一系列高等级精品酒店、写字楼及大型标杆楼盘的设计。公司以"激情打造、致诚服务"的企业精神，为业界提供规划，建筑，室内，园林全方位的设计及咨询服务。

Teamer-arch——Dream works of architectural design. The four dominant leaders have worked in the architectural design for about two decades and been honored to take part in and experience the marvelous development of Chinese real estate industry for the past decade in the first line. Their works can be seen everywhere in the country, especially with unique experiences and profound understandings about the design of massive building clusters and various hotels.

Teamer-arch is a team and we emphasize on the role played by a team and stress on the professional systematic management. The coordination and cooperation is the foundation to guarantee high efficiency and quality.

Teamer-arch is a kind of concept, learning and conceiving from the nature; Teamer-arch is a kind of active spirit, "all going up and can't be stopped."

Teamer Architectural Design and Consultants Co.,Ltd was registered in Guangzhou China. At present, a team of 100 various professional designers have been formed and a series of high-class hotels and massive building clusters have been completed. With "enthusiastic work and sincerest service" as the enterprise spirit, we dedicated to provide planning and design of architecture, internal space, garden and consulting services for the industry.

广州天萌建筑

www.teamer-arch.com

佛山温德姆豪廷酒

广州天萌建筑馆

东林美城

南沙·云山诗意

恩平锦江国际新城

白天鹅（飞来峡）温泉度假酒店

东怡水岸

临沂益华城市综合体

TEAMER ARCH
天萌(中国)建筑设计机构

地址：广州市天河区员村四横路128号红专厂F9栋
TEL：020-37857429
FAX：020-37857590
E-mail：teamer_gz@126.com
http：www.teamer-arch.com

丽湖国际大酒店

 上海中建建築設計院有限公司
SHANGHAI ZHONGJIAN ARCHITECTURAL DESIGN INSTITUTE CO., LTD.

上海中建 · 上海总部	上海中建 · 西安分院
浦东新区东方路989号	南二环西段202号
中达广场12楼	九座花园611室
电话：86-21-6875 8810	电话：86-29-8837 8506

上海中建建筑设计院有限公司，成立于1984年，隶属于中国建筑工程总公司，是国家批准的拥有建筑工程设计甲级、装饰设计甲级、风景园林设计乙级、城市规划编制乙级资质的综合性设计公司。公司是中国勘察设计协会和上海市勘察设计协会会员单位，2011年被授予首批"全国诚信单位"，并入选为中国2010年上海世博会建筑设计类九家推荐服务供应商之一。

公司长期面向海内外开展设计业务，先后与美国、加拿大、法国、比利时、香港等国家和地区的著名设计公司合作设计了多项工程，设计作品遍及国内二十多个省市自治区及海外的俄罗斯、阿尔及利亚等国家，有多项设计作品荣获建设部、上海市及中国建筑工程总公司的各类奖项。

公司将本着敬业、诚信、协作、创新的企业精神，恪守诚信原则，聚焦客户需求，为客户价值的提升呈现专业服务，为人居环境的改善描绘美好蓝图。

上海中建设计

高新区创新国际广场
项目地点：西安高新区
建筑面积：412775 平方米
高度：180米

上海中建·新疆分院
乌鲁木齐市 南湖南路 145 号
上海市政府驻疆办事处 205 室
电 话：86 - 991 - 4161 028

上海中建·济南分院
市中区经四路288号
恒昌大厦 1503 室
电 话：86 - 531 - 80992217

上海中建·安徽分院
合肥市包河区 马鞍山南路
绿地赢海大厦 C座 1110室
电 话：86 - 551 - 3711 566

上海中建·厦门分院
厦门市软件园二期 观日路 601
电 话：86 - 592 - 5999 928

誠聘精英
WWW.SHZJY.COM

BHAD *Architects*
北京奥思得建筑设计有限公司

北京市朝阳区东三环中路39号建外SOHO16号楼2903-2905
TEL：86 10 58692509/2519
Email：bjhad@163.com

呼和浩特东岸国际高尚社区

北京乐府江南居住区

临汾河西企业家总部

北京自主城

BHAD 北京奥思得建筑设计有限公司成立于1994年2月，是中国第一批的中外合资甲级建筑设计公司，并成为英国PCKO建筑事务所、加拿大GBL建筑事务所在中国的代表机构，有着丰富的国际合作设计经验。公司自成立至今承接了国内外几百项大中型重要工程，设计服务质量优良，擅长于高档居住区及别墅区、综合商业建筑、办公建筑、文化建筑、城市设计等众多领域，并涵盖了风景园林、室内设计及机电专项设计等，确立了从项目的前期策划定位、概念研究、方案设计、施工图设计，到施工现场控制指导服务等全方位工作模式，体现出现代设计企业的优秀工作理念，设计作品多次获得国内有关奖项。公司还获评为建设部绿色建筑引导体系建筑设计单位，在业界树立了良好的企业形象。

BHAD北京奥思得建筑设计有限公司集合了众多来自清华大学、天津大学等国内知名院校和海外留学工作归来的专业人士，立志于在中国的城市建设领域做出自己应有的贡献。公司的设计主旨意在将"城市空间、自然环境、个体建筑"三要素作到完美和谐统一，将传统文脉和现代生活作到完美和谐统一。

BHAD北京奥思得建筑设计有限公司近年完成的主要工程包括：北京乐府江南住宅区（北京市优秀社区规划奖）、北京东单正和广场、北京顺驰领海住宅小区、北京自主城居住区、北京百顺达别墅区、北京CBD高尔夫高级公寓、北京银湖别墅、北京芸溪境分时度假村、西宁国际村居住区（建设部优秀规划奖）、上海三林新村住宅区、上海浦东新区高东镇新村住宅区、呼和浩特

包头时代广场

河北邢台万隆广场

葫芦岛北港开发区管委会办公区

阿尔及利亚综合演出厅

临汾奥体中心

东岸国际社区（建设部居住类设计二等奖）、山西大同浩海大厦、包头时代广场、鄂尔多斯左岸公园别墅区、山西浩海大厦（酒店）、山西临汾尧庙国际商贸城、山西临汾奥体中心、温哥华冬奥会运动员村（合作设计）等国内外一系列不同类型的工程，还在国际竞赛中多次荣获奖项，包括：阿尔及利亚布盖迪新城规划（国际竞赛第一名）、阿尔及利亚国家综合演出厅（国际竞赛第二名）等。

BHAD北京奥思得建筑设计有限公司更多地结合国际先进设计和管理理念，既专注于对建筑设计过程中细节的把控，同时十分注重对城市整体发展的研究和结合，将设计作品切实地落实在城市文化高度去思考和实现，具有创新精神地致力于绿色生态建筑和可持续发展的设计研究，努力为国家的建设事业做出自身的贡献。

呼伦贝尔珊瑚墅别墅区

温哥华冬奥会运动员村

衡阳·海通温德姆至尊豪廷大酒店（五星级）

天隐国际

上海天隐建筑设计有限公司

SHANGHAI TIANYIN ARCHITECTURE DESIGN CO;LTD

地址：上海市杨浦区国康路 100 号国际设计中心 1402

网站：www.skyarchdesign.com

电话：021-6598 8000

传真：021-6598 2798

邮编：200092

SJS 四季园林

广州从化. 宝趣玫瑰世界

■ 风景园林专项设计乙级
□ 景观设计　　　　　Landscape Design
□ 旅游建筑设计　　　Tour Architectural Design
□ 旅游度假区规划　　Resorts and Leisure Planning
□ 市政公园规划　　　Park and Green Space Planning

广州市四季园林设计工程有限公司成立于2002年，公司由创始初期从事景观设计，已发展为旅游区规划、度假区规划、度假酒店、旅游建筑、市政公园规划等多类型设计的综合性景观公司。设计与实践相结合，形成了专业的团队和服务机构，诚邀各专业人士加盟。

Add：　广州市天河区龙怡路117号银汇大厦2505
Tell：　020-38273170　　　　Fax：　020-86682658
E-mail:yuangreen@163.com　　Http：//WWW.gz-siji.com

奥森景观公司乔迁新址
深圳市南山区南海大道1061号喜士登大厦四层

　　奥森国际景观设计公司正持续稳步发展壮大,于2012年11月搬迁新址,新办公室属于蛇口网谷,该产业园作为政府打造的战略新兴产业基地,文化氛围浓厚,毗邻大南山,可远观南山山景,环境幽雅,空气清新。办公面积近千平方米,内部环境宽敞明亮,并设有茶水间、阅览室,为公司员工提供了休息放松的空间。奥森国际景观一直致力于成为中国最具有影响力的环境景观规划设计公司,目前的业务范围已涉及风景旅游区、度假别墅区、商务空间、市政景观、综合性公园、城市空间、娱乐空间等景观规划设计。乔迁至新办公地点,奥森景观公司将在新的平台上继续努力,为景观事业贡献自己的绵薄之力。

地址:深圳市南山区南海大道1061号喜士登大厦四层
电话:86755-26828246　　公司邮箱:oc-la@163.com
传真:86755-26822543　　招聘邮箱:oc-hr@163.com
邮编:518000　　　　　　网址:www.oc-la.com

OCEANICA

RESIDENTIAL LANDSCAPE
PUBLIC AREA
INDUSTRY PARK OPEN SPACE
TOURISM ZONE
COMMERCIAL LANDSCAPE
URBAN RENEWAL
HOTELS & RESORTS

前言 EDITOR'S NOTE

筑造城市居住梦想
BUILD THE DREAM OF CITY LIVING

高层公寓往往跟城市紧密相连，伴随城市的不断发展，城市里有限的土地与所承载的高密度人口之间的矛盾日益突出，使得居住建筑考虑竖向发展以争取更多的空间。作为集合式住宅的一种，高层公寓同样可容纳许多的人居住，并且，其作为个人的私密空间在城市中扮演着重要的角色。

高层公寓最突出的特征即建筑造型，无论是采用中轴对称设计、点式布局设计还是圆弧状建筑设计等等，能否与所在城市和谐，使其成为城市结构中的有机组成部分显得至关重要。除了视觉上的美丑，造型与立面的好坏还直接影响公寓的居住品质。因而，设计过程中必然要兼顾立面与户型的设计，创造出美的立面效果的同时营造舒适的居住空间。高层公寓在景观设计方面，更注重俯视时的形式感，强调立体绿化，多采用空中花园等手法。此外，设计中还需考虑到建筑的采光通风、交通流线组织设计以及相关的生活配套设施等等。

高层公寓承载着城市中人们的居住梦想，因而它不仅是人们物质上家的庇护所，同时也是人们心理和精神上的寄托。本期专题围绕高层公寓的设计展开，结合精选的高层公寓经典案例，为您呈现出"家"的设计细节。

The high-rise apartments are often closely linked with the city. With the continuous development of the city, the contradiction between the limited amount of land in the city and the high-density population becomes increasingly prominent. Residential buildings consider developing vertically to obtain more space. As a kind of collective apartment, high-rise apartment can accommodate a good many number of people as well, and it plays an important role in the city as an individual private space.

The most prominent feature of the high-rise apartment buildings is shape, whether in axial symmetry design, point-layout design or arc-shaped architectural design, making it an integral part of the urban structure seems very important. In addition to the visual beauty and ugliness, good or bad shape and facade also directly affect the quality of apartment living. Thus, the design process is bound to take into account the facade and the scale to create a beautiful facade effect and to create a comfortable living space at the same time. In terms of landscape design, high-rise apartments pay more attention to the form when seen from above, highlight 3D greening and create hanging garden mostly. In addition, attention should be paid to lighting, ventilation, traffic flow line organization and relevant living facilities.

High-rise apartment carries the living dream of the city dwellers. Therefore, it is not a material shelter, but also spiritual ballast for people. In this volume, we select some classic works on high-rise apartment to present you the design detail of a real home.

jiatu@foxmail.com

NEWHOUSE 图解地产与设计

2012年 总第45期

面向全国上万家地产商决策层、设计院、建筑商、材料商、专业服务商的精准发行

指导单位 INSTRUCTION UNIT
亚太地产研究中心
中国花卉园艺与园林绿化行业协会

出品人 PUBLISHER
杨小燕 YANG XIAOYAN

主编 CHIEF EDITOR
王志 WANG ZHI

副主编 ASSOCIATE EDITOR
熊冕 XIONG MIAN

编辑记者 EDITOR REPOTERS
唐秋琳 TANG QIULIN
钟梅英 ZHONG MEIYING
胡明俊 HU MINGJUN
康小平 KANG XIAOPING
吴辉 WU HUI
曾伊莎 ZENG YISHA
曹丹莉 CAO DANLI
朱秋敏 ZHU QIUMIN
王盼青 WANG PANQING

设计总监 ART DIRECTORS
杨先周 YANG XIANZHOU
何其梅 HE QIMEI

美术编辑 ART EDITOR
詹婷婷 ZHAN TINGTING

国内推广 DOMESTIC PROMOTION
广州佳图文化传播有限公司

市场总监 MARKET MANAGER
周中一 ZHOU ZHONGYI

市场部 MARKETING DEPARTMENT
方立平 FANG LIPING
熊光 XIONG GUANG
王迎 WANG YING
杨先凤 YANG XIANFENG
熊灿 XIONG CAN
刘佳 LIU JIA
王成林 WANG CHENGLIN

图书在版编目（CIP）数据
新楼盘. 高层公寓：汉英对照 / 佳图文化主编.
—— 北京：中国林业出版社, 2013.1
ISBN 978-7-5038-6949-5

Ⅰ. ①新... Ⅱ. ①佳... Ⅲ. ①建筑设计 - 中国 - 现代 - 图集 Ⅳ. ①TU206

中国版本图书馆CIP数据核字(2012)第 080726 号
出版：中国林业出版社
主编：佳图文化
责任编辑：李顺 许琳
印刷：利丰雅高印刷(深圳)有限公司

特邀顾问专家 SPECIAL EXPERTS (排名不分先后)

赵红红 ZHAO HONGHONG	赵士超 ZHAO SHICHAO
王向荣 WANG XIANGRONG	孙 虎 SUN HU
陈世民 CHEN SHIMIN	梅卫平 MEI WEIPING
陈跃中 CHEN YUEZHONG	林世彤 LIN SHITONG
邓 明 DENG MING	熊 冕 XIONG MIAN
冼剑雄 XIAN JIANXIONG	周 原 ZHOU YUAN
陈宏良 CHEN HONGLIANG	李焯忠 LI ZHUOZHONG
胡海波 HU HAIBO	原帅让 YUAN SHUAIRANG
程大鹏 CHENG DAPENG	王 颖 WANG YING
范 强 FAN QIANG	周 敏 ZHOU MIN
白祖华 BAI ZUHUA	王志强 WANG ZHIQIANG / DAVID BEDJAI
杨承刚 YANG CHENGGANG	陈英梅 CHEN YINGMEI
黄宇奘 HUANG YUZANG	吴应忠 WU YINGZHONG
梅 坚 MEI JIAN	曾繁柏 ZENG FANBO
陈 亮 CHEN LIANG	朱黎青 ZHU LIQING
张 朴 ZHANG PU	曹一勇 CAO YIYONG
盛宇宏 SHENG YUHONG	冀 峰 JI FENG
范文峰 FAN WENFENG	滕赛岚 TENG SAILAN
彭 涛 PENG TAO	王 毅 WANG YI
徐农思 XU NONGSI	陆 强 LU QIANG
田 兵 TIAN BING	徐 峰 XU FENG
曾卫东 ZENG WEIDONG	张奕和 EDWARD Y. ZHANG
马素明 MA SUMING	郑竞晖 ZHENG JINGHUI
仇益国 QIU YIGUO	刘海东 LIU HAIDONG
李宝章 LI BAOZHANG	凌 敏 LING MIN
李方悦 LI FANGYUE	谢锐何 XIE RUIHE
林 毅 LIN YI	姜 圣 JIANG SHENG
陈 航 CHEN HANG	章 强 ZHANG QIANG
范 勇 FAN YONG	

编辑部地址: 广州市海珠区新港西路3号银华大厦4楼
电话: 020—89090386/42/49、28905912
传真: 020—89091650

北京办: 王府井大街277号好友写字楼2416
电话: 010—65266908　　**传真:** 010—65266908

深圳办: 深圳市福田区彩田路彩福大厦B座23F
电话: 0755—83592526　　**传真:** 0755—83592536

协办单位 CO—ORGANIZER

 广州市金冕建筑设计有限公司　熊冕 总设计师
地址：广州市天河区珠江西路5号国际金融中心主塔21楼06—08单元
TEL：020—88832190　88832191
http://www.kingmade.com

AECF 上海颐朗建筑设计咨询有限公司　巴学天 上海区总经理
地址：上海市杨浦区大连路970号1308室
TEL：021—65909515　　FAX：021—65909526
http://www.yl—aecf.com

WEBSITE COOPERATION MEDIA
网站合作媒体

SouFun 搜房网

副理事长单位 DEPUTY CHAIRMAN

华森建筑与工程设计顾问有限公司 邓明 广州公司总经理
地址：深圳市南山区滨海之窗办公楼6层
　　　广州市越秀区德政北路538号达信大厦26楼
TEL：0755—86126888　020—83276688
http://www.huasen.com.cn　E—mail:hsgzaa@21cn.net

上海中建建筑设计院有限公司 徐峰 董事长
地址：上海市浦东新区东方路989号中达广场12楼
TEL：021—68758810　FAX：021—68758813
http://www.shzjy.com
E—mail：csaa@shzjy.com

广州瀚华建筑设计有限公司 冼剑雄 董事长
地址：广州市天河区黄埔大道中311号羊城创意产业园2—21栋
TEL：020—38031268　FAX：020—38031269
http://www.hanhua.cn
E—mail：hanhua—design@21cn.net

常务理事单位 EXECUTIVE DIRECTOR OF UNIT

深圳市华域普风设计有限公司 梅坚 执行董事
地址：深圳市南山区海德三道海岸城东座1306—1310
TEL：0755—86290985　FAX：0755—86290409
http://www.pofart.com

天萌（中国）建筑设计机构 陈宏良 总建筑师
地址：广州市天河区员村四横路128号红专厂F9栋天萌建筑馆
TEL：020—37857429　FAX：020—37857590
http://www.teamer—arch.com

天友建筑设计股份有限公司 马素明 总建筑师
地址：北京市海淀区西四环北路158号慧科大厦7F（方案中心）
TEL：010—88592005　FAX：010—88229435
http://www.tenio.com

奥雅设计集团 李宝章 首席设计师
深圳总部地址：深圳蛇口南海意库5栋302
TEL：0755—26826690　FAX：0755—26826694
http://www.aoya—hk.com

广州山水比德景观设计有限公司 孙虎 董事总经理兼首席设计师
地址：广州市天河区珠江新城临江大道685号红专厂F19
TEL：020—37039822/823/825　FAX：020—37039770
http://www.gz—spi.com

广州市四季园林设计工程有限公司 原帅让 总经理兼设计总监
地址：广州市天河区龙怡路117号银汇大厦2505
TEL：020—38273170　FAX：020—86682658
http://www.gz—siji.com

深圳市佰邦建筑设计顾问有限公司 迟春儒 总经理
地址：深圳市南山区兴工路8号美年广场1栋804
TEL：0755—86229594　FAX：0755—86229559
http://www.pba—arch.com

北京博地澜屋建筑规划设计有限公司 曹一勇 总设计师
地址：北京市海淀区中关村南大街31号神舟大厦8层
TEL：010—68118690　FAX：010—68118691
http://www.buildinglife.com.cn

香港华艺设计顾问（深圳）有限公司 林毅 总建筑师
地址：深圳市福田区华富路航都大厦14、15楼
TEL：0755—83790262　FAX：0755—83790289
http://www.huayidesign.com

华通设计顾问工程有限公司
地址：北京市西城区西直门南小街135号西派国际C—Park3号楼
TEL：8610—83957395　8610—83957390
http://www.wdce.com.cn

GVL国际怡境景观设计有限公司 彭涛 中国区董事及设计总监
地址：广州市珠江新城华夏路49号津滨腾越大厦南塔8楼
TEL：020—87690558　FAX：020—87697706
http://www.greenview.com.cn

R—LAND 北京源树景观规划设计事务所 白祖华 所长
地址：北京朝阳区朝外大街怡景园5—9B
TEL：010—85626992/3　FAX：010—85625520
http://www.ys—chn.com

北京寰亚国际建筑设计有限公司 赵士超 董事长
地址：北京市朝阳区琨莎中心1号楼1701
TEL：010—65797775　FAX：010—84682075
http://www.hygjjz.com

奥森国际景观规划设计有限公司 李焯忠 董事长
地址：深圳市南山区南海大道1061号喜士登大厦四楼
TEL：0755—26828246　86275795　FAX：0755—26822543
http://www.oc—la.com

深圳市雅蓝图景观工程设计有限公司 周敏 设计董事
地址：深圳市南山区南海大道2009号新能源大厦A座6D
TEL：0755—26650631/26650632　FAX：0755—26650623
http://www.yalantu.com

北京新纪元建筑工程设计有限公司 曾繁柏 董事长
地址：北京市海淀区小马厂6号华天大厦20层
TEL：010—63483388　FAX：010—63265003
http://www.bjxinjiyuan.com

HPA上海海波建筑师事务所 陈立波、吴海青 公司合伙人
地址：上海市中山西路1279弄62号楼国峰科技大厦11层
TEL：021—51168290　FAX：021—51168240
http://www.hpa.cn

哲思（广州）建筑设计咨询有限公司 郑竞晖 总经理
地址：广州市天河区天河北路626号保利中宇广场A栋1001
TEL：020—38823593　FAX：020—38823598
http://www.zenx.com.au

理事单位 COUNCIL MEMBERS （排名不分先后）

广州市柏澳景观设计有限公司 徐农思 总经理
地址：广州市天河区广园东路2191号时代新世界中心南塔2704室
TEL：020—87569202　FAX：020—87635002
http://www.bacdesign.com.cn

北京奥思得建筑设计有限公司 杨承冈 董事总经理
地址：北京朝阳区东三环中路39号建外SOHO16号楼2903~2905
TEL：86—10—58692509/19/39　FAX：86—10—58692523

广州嘉柯园林景观有限公司 陈航 执行董事
地址：广州市珠江新城华夏路49号津滨腾越大厦北塔506—507座
TEL：020—38032521/23　FAX：020—38032679
http://www.jacc—hk.com

CDG国际设计机构 林世彤 董事长
地址：北京海淀区长春路11号万柳亿城中心A座10/13层
TEL：010—58815603　58815633　FAX：010—58815637
http://www.cdgcanada.com

上海唯美景观工程有限公司 朱黎青 董事、总经理
地址：上海市徐虹中路20号2—202室
TEL：021—61122209　FAX：021—61139033
http://www.wemechina.com

深圳灵顿建筑景观设计有限公司 刘海东 董事长
地址：深圳福田区红荔路花卉世界313号
TEL：0755—86210770　FAX：0755—86210772
http://www.szld2005.com

广州邦景园林绿化设计有限公司 谢悦何 董事及设计总监
地址：广州市天河北路175号祥龙花园晖祥阁2504/05
TEL：020—87510037　020—38468069
http://www.bonjinglandscape.com

中房集团建筑设计有限公司 范强 总经理／总建筑师
地址：北京市海淀区百万庄建设部院内
TEL：010—68347818

陈世民建筑师事务所有限公司 陈世民 董事长
地址：深圳市福田中心区益田路4068号卓越时代广场4楼
TEL：0755—88262516/429

侨恩国际（美国）建筑设计咨询有限公司
地址：重庆市渝北区龙湖MOCO4栋20—5
TEL：023—88197325　FAX：023—88197323
http://www.jnc—china.com

广州市圆美环境艺术设计有限公司 陈英梅 设计总监
地址：广州市海珠区宝岗大道杏坛大街56号二层之五
TEL：020—34267226　83628481　FAX：020—34267226
http://www.gzyuanmei.com

上海金创源建筑设计事务所有限公司 王毅 总建筑师
地址：上海杨浦区黄兴路1858号701—703室
TEL：021—55062106　FAX：021—55062106—807
http://www.odci.com.cn

深圳市奥德景观规划设计有限公司 凌敏 董事总经理、首席设计师
地址：深圳市南山区蛇口海上世界南海意库2栋410#
TEL：0755—86270761　FAX：0755—86270762
http://www.lucas—designgroup.com

上海天隐建筑设计有限公司 陈锐 执行董事
地址：上海市杨浦区国康路100号国际设计中心1402室
TEL：021—65988000　FAX：021—65982798
http://www.skyarchdesign.com

目录 CONTENTS

013 前言 EDITOR'S NOTE

018 资讯 INFORMATION

名家名盘 MASTER AND MASTERPIECE

022 保利西塘越：经典海派风格建筑 首席人文养老住区
CLASSICAL SHANGHAI STYLE ARCHITECTURE FIRST-CLASS HUMANE ENDOWMENT COMMUNITY

028 南通和融优山美地•名邸：新古典法式格调的全龄化社区
ALL-AGED COMMUNITY WITH NEO-CLASSICAL FRENCH STYLE

034 太仓绿地新城：沉稳 大气 典雅 精致
PROFOUND, AMBIENT, ELEGANT AND EXQUISITE

专访 INTERVIEW

040 专一才能创造独特价值
——访安道（香港）景观与建筑设计有限公司设计运营经理 宋晟
ONLY CONCENTRATION CAN CREATE UNIQUE VALUE

新景观 NEW LANDSCAPE

044 公寓大楼：强调立体感与通透性的生态化空间
3D ECOLOGICAL SPACE WITH TRANSPARENCY

050 上海古北佘山别墅景观设计：流动的音乐 梦想的生活
FLOATING MUSIC, DREAMY LIFE

专题 FEATURE

060 武汉万科金色家园：错跃的表皮结构 轻松动感的空间
RANDOM SKIN AND DYNAMIC SPACE

066	印度尼西亚雅加达豪华公寓： 极富动感与现代元素的帆船状建筑 DYNAMIC MODERN SAIL-SHAPED BUILIDING
070	空中公寓：多元化立面构造下的被动式住宅 PASSIVE RESIDENCE WITH DIVERSIFIED ELEVATIONS
078	大都会：融实用性与舒适性于一体的国际化住区 PRACTICAL AND COMFORTABLE UPSCALE COMPLEX

新特色 NEW CHARACTERISTICS

084	大连金州鸿玮澜山一期：沉稳 优雅的北美风情建筑 COMPOSED AND ELEGANT NORTH AMERICAN ARCHITECTURE
092	上海绿地新都会：现代简约的法式风格品质社区 MODERN, ELEGANT AND HIGH-QUALITY FRENCH-STYLE COMMUNITY

066

098	六盘水未来之城：生态、典雅的新古典主义城市社区 ECOLOGICAL AND ELEGANT NEOCLASSICAL URBAN COMMUNITY

新空间 NEW SPACE

104	海洋中心：现代阁楼生活 MODERN LOFT LIVING

新创意 NEW IDEA

108	巴黎篮子学生公寓：立面丰富多变 极具活力的建筑 DYNAMIC BUILDING WITH DIVERSIFIED FACADES
118	阿姆斯特丹斯卡拉大型住宅楼： 立面丰富、表皮多变的精致砖结构建筑 DETAILED BRICK ARCHITECTURE WITH NUANCED SKIN

084

118

商业地产
COMMERCIAL BUILDINGS

126	阿姆斯特丹MINT酒店：棱角突出的传统荷兰风格建筑 ON THE EDGE OF TRADITIONAL DUTCH STYLE ARCHITECTURE
134	泰州万达广场：简洁现代、完整大气的新一代城市综合体 CONCISENESS AND MODERN, INTEGRITY AND ELEGANCE OF A NEW GENERATION OF URBAN COMPLEX
140	FDA总部：互动 高效 通达 简练 INTERACTION HIGH PERFORMANCE MASTERY CONCISE
146	成都国际财智科技产业园区：古典、庄重、严谨的创意性空间 CLASSICAL, SOLEMN, COMPACT AND INNOVATIVE SPACE

126

INFORMATION | 资讯/地产

中央政治局：明年要加强房地产调控

中共中央政治局12月4日召开会议，分析研究2013年经济工作。会议提出，要保持宏观经济政策的连续性和稳定性，着力提高针对性和有效性，适时适度进行预调微调，加强政策协调配合。会议提出，要积极稳妥推进城镇化，增强城镇综合承载能力，加强房地产市场调控和住房保障工作。

THE CENTRAL POLITICAL BUREAU: NEXT YEAR TO ENHANCE THE REAL ESTATE REGULATION

On Dec. 4, the Political Bureau of CPC Central Committee held a meeting, analysis of the 2013 economic work, and pointed the need to implement a proactive fiscal policy and prudent monetary policy, and enhance the relevance, flexibility and effectiveness of macro-control. It pointed out the need to promote urbanization, enhance the carrying capacity of cities and towns, and continue real estate market regulation and low-income housing projects.

统计局：1~11月全国房地产投资64 772亿元

12月9日，国家统计局公布今年1~11月份全国房地产开发和销售情况数据。2012年1~11月份，全国房地产开发投资64 772亿元，同比名义增长16.7%，增速比1~10月份提高1.3个百分点。其中，住宅投资44 606亿元，增长11.9%，增速提高1.1个百分点，占房地产开发投资的比重为68.9%。

NATIONAL BUREAU OF STATISTICS OF CHINA: NATIONAL REAL ESTATE INVESTMENT 6,477.2 BILLION YUAN FROM JANUARY TO NOVEMBER

On Dec. 9, National Bureau of Statistics of China announced the data for national real estate development and sales from January to November. The total investment in real estate development from Jan. to Nov. was 6,477.2 billion yuan, a year-one-year growth of 16.7 percent, increased 1.3 percentage points over the first ten months. Of which, the investment in residential buildings was 4,460.6 billion yuan, up by 11.9 percent, 1.1 percentage points higher than that in the first ten months, and accounted for 68.9 percent of real estate development investment.

财政部部长：要逐步在全国推开房产税

财政部部长谢旭人11月21日在其署名文章中表示，要统筹推进房地产税费改革，逐步改变目前房地产开发、流转、保有环节各类收费和税收并存的状况。认真总结个人住房房产税改革试点经验，研究逐步在全国推开，同时积极推进单位房产的房产税改革。

MINISTER OF FINANCE:STEADILY PUSH FORWARD PROPERTY TAX REFORM TRIALS

Chinese Finance Minister Xie Xuren said in a signed article on Nov. 21 that, China is mulling further reforms of property tax and expansion of property tax trials, gradually changing the coexistence of fees and taxations on property development, transactions and possessions. There is a need to summarize the experience in private property tax reform pilots and steadily expand it all through the country. At the same time, it needs to push forward the property tax reform in institutions' properties.

"绿标．LEED．绿色建筑新趋势" 主题论坛成功举办

2012年11月30日下午，由中国绿色建筑与节能（香港）委员会携梁黄顾建筑师（香港）事务所有限公司在深圳华侨城华会所成功举办了"绿标．LEED．绿色建筑新趋势"主题论坛。众多来自政府机关、学术团体、中港两地建筑协会、建筑设计及相关行业的专家和代表参加了这次活动，论坛气氛热烈，各界代表积极探讨了绿标与LEED在中国的发展及相关案例，为探索中国绿色建筑的未来发展方向，进一步发展绿色建筑设计开拓了新的思路，也为建筑界可持续发展战略提供了参考和借鉴。

FORUM "GREEN BUILDING LABEL, LEED, NEW TREND OF SUSTAINABLE BUILDING" SUCCEEDED

On the afternoon of Nov. 30, 2012, China Green Building (Hong Kong) Council, "CGBC(HK)" for short, together with LWK & Partners (HK) Ltd., had held the forum "Green Standard 'LEED' – New Tendency for Green Building" in OCT Shenzhen Clubhouse successfully. Many experts and representatives from governments, research groups, architectural associates of mainland and Hong Kong, architectural institutes, and the relative industries had participated in this forum, discussing and studying on the development of green building label and LEED in China. It is trying to explore the future development for sustainable buildings, bringing new ideas for sustainable building design, and providing references for the sustainable development in building industry.

深圳年度"地王"诞生

11月28日,深圳尖岗山一块面积达15万m²的土地出让,起价10亿元,最终中海地产以20亿元成功竞标获得该地块。总价20亿元使得这一地块成为深圳今年新的总价地王。同时,竞得这一地块的中海地产,按约定还将承担建设深圳当代艺术馆与城市规划展览馆,并运营管理20年。

SHENZHEN'S MOST EXPENSIVE LAND BORN

On Nov. 28, a 150,000㎡ plot in Jiangangshan of Shenzhen was sold by auction with a initial bidding price of 1 billion yuan. COHL finally won the bid with a total amount of 2 billion yuan, which makes the new most-expensive-land of Shenzhen come into being. According to the agreement, the bid winner COHL will also be in charge of the construction and 20-year management of Shenzhen Modern Art Museum and Urban Planning Exhibition Hall.

万科广州白云新城项目将建13栋住宅

广州市规划局于12月8日挂出白云新城万科项目批后公示,该项目拟建10栋17层、3栋7层住宅,辅以两栋两层餐饮以及一层公建配套。据了解,项目所在地块由万科于2011年以12.7亿元摘得,折合楼面价13 536元/平方米,预计建成日期为2014年6月。

VANKE BAIYUN NEW TOWN TO BUILD 13 RESIDENTIAL BUILDINGS

According to the administrative approval by Guangzhou Planning Bureau on December 8, Baiyun New Town will include ten 17-storey and three 7-storey residential buildings, complemented with two 2-storey facilities for catering, and one floor for public services. It is known that Vanke won this land in 2011 with a total amount of 1.27 billion yuan, averaging 13,536 yuan a square meter. It is expected to be realized in June, 2014.

"上海中心"高度破400米

据报道,12月8日上午,上海中心大厦主楼核心筒第86层楼板浇筑完成,成功突破400米,达到403.2米。预计到12月底,可达到424米,高度将超过毗邻的金茂大厦。报道称,上海中心即将进入商业招商阶段。目前其商务社区总体功能空间规划已确定,并已成立了商务运营公司负责市场开发招商运营。

SHANGHAI TOWER RISES OVER 400 METERS

It is reported that, on the morning of Dec. 8, Shanghai Tower, a skyscraper under construction, had its 86th floor casted. Till then the tower reached 403.2 meters. It is expected to reach 424 meters by the end of December, which will surpass the neighboring Jin Mao Tower. It is said that Shanghai Tower is about to attract investments. At present, the overall layout of the functional spaces in the business area is decided and it has set the management company to develop the market.

时代地产八亿珠海斗门揽地

12月6日,珠海市以公开挂牌方式出让宗地编号为珠国土储2012-14号地块的国有建设用地使用权,经过多轮竞价,最终由珠海市佳誉房地产开发有限公司以8.36亿元夺得。据悉,该公司是时代地产布局珠江口西岸的重点城市公司。据了解,斗门区是珠海未来城区发展规划轴心,而白蕉将成为斗门的新中心。白蕉片区基础设施良好,位置适中,并且已经有万科、华发、三一集团等品牌开发商进驻。

旭辉集团首11月销售85.4亿 比去年全年高57%

旭辉集团12月10日宣布,于2012年1至11月,集团累计实现合同销售金额人民币约85.4亿元,合同销售面积约93.2万m²。合同销售均价约人民币9 160元/平方米。旭辉集团称,首11个月合同销售金额和合同销售面积分别超过2011年全年合同销售金额和合同销售面积的57%及72%,亦已超过本年度全年的销售目标。

GIFI GROUP: 11 MONTHS' SALE REACHES 8.54 BILLION YUAN, 157% OF 2011'S WHOLE-YEAR SALE

On Dec. 10, GIFI Group announced that from Jan. to Nov., its total sales amount reaches 8.54 billion yuan and the total sold area is 932,000 m², averaging 9,160 yuan per square meter. It is said that the 11 months' sale amount is 157% of the last year's, and the contracted sold area is 72% more than 2011. 2012 sales targets are fulfilled ahead of schedule.

TIMES PROPERTY: 0.836 BILLION YUAN FOR A PLOT

On Dec. 6, plot No. 2012-14 of Zhuhai was sold at a land auction. After serval rounds of biddings, the winning bidder Zhuhai Jiayu Real Estate Development Co., Ltd. paid 0.836 billion yuan for this plot. It is known that Jiayu is an important subsidiary of Times Property on the west bank of Pearl River. Doumen District is the core area in the future Zhuhai City, and Baijiao area will be the new downtown of Doumen. Baijiao is ideally located with favorable municipal facilities. There have already been many developments by Chinese real estate giants such as Vanke, Huafa, Sany Group and so on.

INFORMATION 资讯/设计

卡布雷拉公寓

用地周围有很多大学校园，而学生公寓在该地区非常紧缺。这座建筑就是为了满足学生公寓的需要而建。在这座建筑中，设计师共设计了五种户型：工作室、共享客厅的两居、独立两居、传统户型和复式住宅。并在第九层设计了多种不同的公共功能空间。

Cabrera

The project was created in response to the high demand for a student housing, a need that arises from the location of the building in an area of the city with many universities. The architects have developed five types of housings, providing a high level of independency in their internal organization. They are: Studio Apartments; two rooms with square common areas; two divided rooms; two traditional rooms; duplex. The architects proposed a variety of programs that promoted the unity and integrity of people, which are located on the ninth level.

Ber住宅

Ber住宅是Nico van der Meulen Architects设计事务所与M Square Lifestyle Design最近合作设计的一个项目，是将花岗岩、钢构件、光与水组合在一起的结果。这座建筑位于南非的米德兰，其最大特点是不规则的铁栏杆随机排列出的图案。

House Ber

House Ber, the latest work by Nico van der Meulen Architects and M Square Lifestyle Design is an indication of what happens when granite, steel, light and water come together. Situated in Midrand, House Ber presents itself as a sequence of irregular steel bars randomly placed creating patterned facades.

斜坡上的木屋

该用地原本是设置缆车用地的一部分，现在依山就势坐落在那里的是一座全部用木板装饰的小木屋，由瑞士设计工作室studio lx1设计。木屋各个方向都设置了不同大小的窗，可以看见东方的森林和裸露的岩石，西方的雪山和北方860 m以下被雨水冲刷的山谷。室内采用错层式的空间组织形式，不但顺应了斜坡地形，还很好的保护了所有房间的私密性。

Wooden House on the Slope

as part of a site originally intended for the realization of a cable car, a monolithic wooden object plays with the topography and the context. designed by lausanne-based architecture studio lx1, the wooden facade surfaces react to the building by deforming an exalt to the surrounding views, making visual connections in all directions: forest and outcropping rock to the east, snowy peaks to the west, and rhone valley to the north, 860 meters below. using a split-level organization, the building incorporates the slope and ensures the privacy of the various parts of the program: main apartment: day – night spaces, secondary apartment, medical massage office and workshop.

巴塞罗那保障房

该项目包括5种功能类型：为年轻人和老人准备的154套公租房，一个托儿所，一个社区教育中心和两个地下出租停车场。计划中的所有功能都组织在两座独特的建筑中，被一个中心庭院分开，位于两条繁华街道交叉口的一角，位置非常明显。

154 Rental Social Housing and Public Building

Five projects for a collection of public use buildings: 154 rental dwellings for young people and senior citizens; a nursery; a community centre; and two underground storeys for rental parking lots. The scheme is organized into two distinct buildings separated by a courtyard located in the most prominent position.

梧桐住宅

这座梧桐住宅是Kovac Architects设计事务所为本公司的负责人Michael Kovac设计的，是目前公司正在进行的节能建筑研究的实验室，同时也展示着Kovac典型的设计哲学。这座建筑完全的融入了周围自然环境，采用了环保建筑材料，使得这座建筑成为加利福尼亚第一座获得美国绿色建筑白金奖的住宅。

Sycamore House

Designed as the home of firm Principal Michael Kovac, Sycamore House serves as a laboratory for the firm's ongoing research into sustainable architecture and a showcase for Kovac's design philosophy. The home's seamless integration of environmental systems and green materials has made it one of the first in California to garner Platinum Certification from the USGBC LEED for Homes Program.

H住宅

H住宅用地位于一座山脚下，周围都是低矮的建筑。屋顶从后向前逐渐降低，水边的餐厅区屋檐是最低的地方。上层的阁楼可以被分为两个内庭，靠近餐厅区，此外，餐厅和卧室间设置了窗洞，这样无论家人们分散在哪些房间，他们彼此间的联系都依然紧密。

House H

The site is in the corner of the hill lot, surrounded of low-rise buildings. The roof is set in accordance with the shape that falls gently toward the back, placed lower is the dining room in the level of the water surroundings. The upper loft space that can be used to divide into two chambers, the future layout will be near from dining. In addition, the installation of windows that opens to the dining and bedroom vanity. With these operations the family can always stay with each other part of the family no matter in which part of the house you are.

集装箱星巴克

这座星巴克汽车餐厅和步行店位于西雅图郊外，由四个翻新的船运集装箱构成。这座建筑的设计理念是受到该公司海外进口咖啡和茶叶的影响，从全球各处运来的商品都是经过这样的集装箱来到这里的。

Starbucks Drive-thru and Walk-up store
Tukwila, a suburb of seattle, is home to Starbucks' reclamation drive-thru and walk-up store, the coffeehouse giant's first space made from four refurbished shipping containers. The concept behind the design was influenced by the company imports of the coffee and tea goods from abroad, all transported from across the globe in such cargo containers.

雷纳家具店

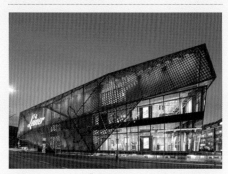

如果通过A12高速路从东面进入因斯布鲁克，从很远的地方就会看到这座新建的雷纳家具店。这座闪耀着该公司标志性绿色的建筑对路人来说就像一座灯塔一样夺目。除了色彩原因以外，建筑本身带棱角的形体也给人留下了深刻的印象。透明的金属网表皮像折纸一样将建筑罩在里面。

Leiner Furniture Store
The new Leiner furniture store can be seen from a considerable distance as you approach Innsbruck from the East on the A12 motorway. The four story structure is a beacon, in Leiner's corporate green, to people passing by. Along with the large areas of green, the angular form of the building itself makes a strong impression. Like a folded tablecloth, the translucent metal construction envelopes the volumes of the building.

阿波座市政中心

土库曼斯坦政府最近委托RTA-Office在阿波座旅游区总体规划范围内设计了一个国会大厅和会议中心。该项目将会向这一地区注入活力和创新精神，同时也会成为一个举行会议并广泛交换信息的地方。

Awaza Congress Hall and Convention Centre
Turkmenistan's government recently commissioned RTA-Office to design a major Congress Hall and Convention center in the Awaza Touristic Region Masterplan. The project will give to the area an injection of life, creativity and innovation, as well as to become a place for meeting and exchange of information in a larger scale.

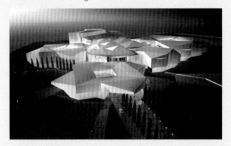

福克旺图书馆

这座建筑位于德国艾森，是福克旺艺术大学的音乐图书馆，由瑞士设计师Max Dudler设计。原先这座巴洛克风格的修道院既是僧人的住所也充当监狱，后来改建为学校，图书馆的所在地原来是军事医院，后来被损毁。设计师想要在原基础上创造一个浑然一体的、陈旧粗糙的石墙，与广场周围的其他建筑一样，前面采用大台阶与广场相连。

Folkwang Library
This music library at Folkwang University of the Arts located in Essen, Germany is designed by Swiss architect Max Dudler. The grand baroque buildings of the former abbey had served as both a residence and a prison before their conversion to a university and the library sits over the foundations of a previously demolished military hospital. Dudler's concept for the new library was to create "a monolithic body built atop the level base of an old rough stone wall", with a stepped entrance from the plaza that would reference the raised approaches of the neighbouring buildings.

兰溪庭

兰溪亭位于成都的国际非物质文化遗产公园，由三个部分组成：餐厅、内部庭院和私人俱乐部。这座建筑通过数字技术对中国传统建筑进行了全新的演绎。建筑的空间布局也是对中国传统南方庭院的全新诠释。

The Lanxi Curtilage
The Lanxi Curtilage is located at the International Intangible Cultural Heritage Park in Chengdu, China. It is composed of three parts: restaurant, inner courtyard and private club. It is an interpretation of traditional Chinese architecture through the language of digital fabrication methods.The spatial layout of this project represents a new interpretation of a traditional South China Garden.

中联重科总部

加利福尼亚设计公司amphibianArc受宇宙飞船形象的启发，为中国机械公司设计出两座并立但不相同的摩天楼，地点位于湖南长沙。这个项目的业主是中国工业车辆制造商中联重科，曾经受托采用amphibianArc先前设计的可变形的展览中心来进行一些公司内部的展览和产品展示。

Zoomlion Headquarters International Plaza
Californian firm amphibianArc was inspired by images of spacecrafts to come up with two different skyscraper proposals for the headquarters of a Chinese machinery company in Changsha.The designs, for industrial vehicle manufacturer Zoomlion, were commissioned following amphibianArc's previous proposals for a shape-shifting exhibition centre to host the company's exhibitions and product displays.

CLASSICAL SHANGHAI STYLE ARCHITECTURE FIRST-CLASS HUMANE ENDOWMENT COMMUNITY

| Jiashan Poly Xitangyue

经典海派风格建筑 首席人文养老住区 —— 保利西塘越

项目地点：中国浙江省嘉善县
开 发 商：上海保利房地产开发有限公司
建筑设计：上海霍普建筑设计事务所有限公司
建筑规模：282 000 m²

Location: Jiashan County, Zhejiang, China
Developer: Poly Shanghai Real Estate Development Co., Ltd.
Architectural Design: Shanghai Hoop Architecture Design Co., Ltd.
Project Area: 282,000 m²

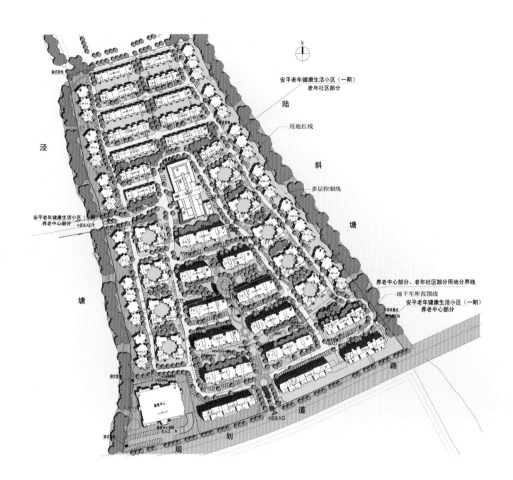

总平面图 Site Plan

鸟瞰图 Aerial View

MASTER AND MASTERPIECE | 名家名盘

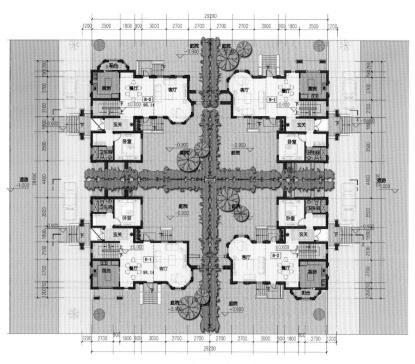

一层平面图 First Floor Plan

项目概况

保利西塘越地处嘉善千年古镇西塘，基地三面环水。总建筑规模为282 000 m²，其中一期占地面积99 720 m²，一期建筑面积为125 120 m²。产品包含老年住宅、养生会所、老年康复中心和理疗中心，结合旅游医疗和居家养生为一体。西塘越将昔日老上海的万千风情融入西塘，以浓厚地域特色的传统文化为根基，用国际时尚的海派风情勾勒出西塘的新天地。

规划布局

根据本地块的三面环水的特点和特殊地理位置，结合本小区养老与基本社区相结合的设计定位，在规划上考虑更好的利用与保护自然，传承独特的江南气质和海派风格，以及对水资源的充分挖掘。

住宅布局以环状模式，外低内高。亲水住宅在最外圈，与河岸毗邻，共39户，合院住宅为中圈，共36户，联排住宅位于内圈，共77户。小高层区包括普通住宅348

户、老年公寓176户、辅老公寓48户。小高层区位于基地的核心区域，住宅部分布局使得更多的住户可以享受到得天独厚的河景资源，基地中心处为养身会所。老年康复中心位于南侧一期小区外围，与项目二期建成的老年医疗中心共同形成整个项目新的核心，并形成第二主入口，服务于周围的老年公寓和辅老公寓，沿河绿化带进行重新整治并向公众开放。

建筑设计

保利西塘越的石库门建筑风格让人耳目一新，其设计灵感来源于一百多年前上海租界中西合璧的海派建筑，并汲取现代建筑设计元素，外观高贵浪漫，富含文化底蕴，可谓独树一帜。保利西塘越高端产品质素必带来高端圈层的汇聚，代表的尊贵感不言而喻。为了能与整个西塘氛围相融合，小高层部分不仅使用灰砖黑瓦，同时出现更多的白色元素，并且在细节处理上向西塘古镇的元素靠拢。康复中心和养生会所则以新中式的白墙灰瓦出现，在建筑群中成为亮点，更加拉近西塘越与西塘古镇的距离。

景观设计

基地为自然形成的三角形河中岛屿，水乡气质展露无疑，自然环境优美，项目依托天然河道，充分利用优越的自然资源，便捷的交通与天然的环境优势，使得保利西塘越得以成为嘉善第一养生豪宅。

MASTER AND MASTERPIECE | 名家名盘

Profile

The project is located in the one thousand years old town–Xitang in Jiashan County, with three sides facing water. It covers an area of 282,000 m² with 99,727 m² of the first phase and 125,120 m² of the second phase. The project includes endowment residence, health–preservation club, senior health recovery center and physical therapy center. Meanwhile, tourist medical care and home health care are also combined in this project. Xitangyue mixes the flirtatious expressions of old Shanghai into Xitang, draws out the sketch of New Xitang of international fashionable styles with intense and profound traditional cultural features.

Planning and Layout

Based on the water–embracing environment, special geographic location and the original design concept of endowment community, the project takes more consideration on natural environment protection in planning to inherit the distinctive Southern temperament and Shanghai style and the fully excavation of water resources as well.

The residential layout is in consistent with environment, with higher levels inside and lower levels outside. Water-intimate residences are placed on the outer ring, closed next to the riverbank, with 39 apartments in total. Courtyard residences are in the middle ring, 36 apartments in total while townhouse residences are arranged in the inner ring with 77 apartments. The intermediate height residential buildings are in the very core of the community, possessing 348 normal apartments, 176 endowment apartments and 48 auxiliary apartments. The arrangement scheme tries to present the river resources to as many residences as possible. A health-preservation club is built right in the center of the community. The aged recovery center is situated outside the first phase in the south, forming a second entrance, as well as the key area of the community with the aged medical center built with the second phase. The endowment apartments and auxiliary apartments which provide service for neighborhood are readjusted along the greening belt and ready to open to the public.

Architectural Design
The stone gate of Poly Xitangyue presents a brand new image, whose design inspiration comes from the combination of Chinese and Western elements—Shanghai style in Shanghai Concession over 100 years ago. Meanwhile, the design also adopts modern architectural design elements with dignified and romantic appearance and profound cultural connotation, special and unique. The high-end product quality of Poly Xitangyue will surely attract the congregation of high-end level, displaying superior dignity. In order to blend with the surrounding Xitang atmosphere, the intermediate height residential buildings adopt not only grey tiles and black tiles, but also more white elements and try to come closer toward Xitang's old town style. The recovery center and health-preservation club are dressed in white walls and grey tiles, which become the bright spots in the building cluster and drag closer the distance with Xitangyue.

Landscape Design
The site is a natural triangle island in a river, which presents apparent waterside temperament with beautiful natural environment. With natural river passage and superior natural resources, the project enjoys convenient transportation and pleasant environment, which makes Poly Xitangyue become the top 1 health-preservation housing.

ALL-AGED COMMUNITY WITH NEO-CLASSICAL FRENCH STYLE

Herong Renowned Mansion with Superior Mountains and Landscape, Nantong

新古典法式格调的全龄化社区 —— 南通和融优山美地·名邸

项目地点：中国江苏省南通市
开 发 商：南通融邦房地产开发有限公司
规划设计：上海霍普建筑设计事务所有限公司
建筑设计：上海霍普建筑设计事务所有限公司
　　　　　南通市建筑设计研究院有限公司
景观设计：贝尔高林国际有限公司
建筑规模：480 000 m²

Location: Nantong, Jiangsu, China
Developer: Nantong Rongbang Real Estate Development Co., Ltd.
Planning: Shanghai Hoop Architectural Design Consultant Inc
Architectural Design: Shanghai Hoop Architectural Design Consultant Inc
　　　　　　　　　　Nantong Institute of Architectural Design Co., Ltd
Landscape Design: Belt Collins
Area: 480,000 m²

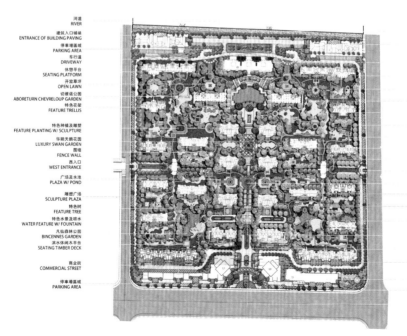

景观总平面图
Master Landscape Plan

MASTER AND MASTERPIECE | 名家名盘

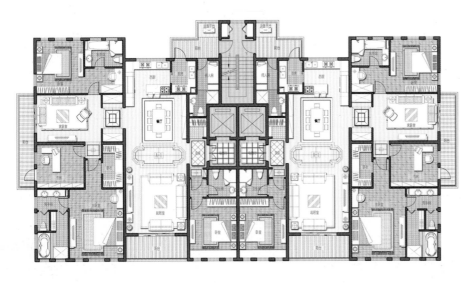

D户型标准层平面图
Standard Floor of Type D

项目概况

项目坐落于南通市政府重点打造的开发区核心区内，南临振兴东路、东临新景路、西靠新开河。距南通市中心约10 km。项目毗邻著名的狼山风景区及星湖101商圈，定位为中高端法式全龄化住区，规划有高层住宅、配套商业和托幼设施，交通便捷且周边配套齐全。

规划布局

在整体布局上以经典的法式中轴对称理念为核心，在古典的建筑形式和异域风情的景观格调基础上巧妙地融入了现代元素和中式园林造景手法。采用小组团化邻里社区，将地块自然划分为四个区块，建筑排布顺应地势景观以及朝向的变化，在各个区块内形成意趣盎然的空间形态。并在此基础上，将景观加以优化，

创造更加丰富的空间层次、步移景异、形态自然的人居环境，为居住者赢取空间、建筑、景观与生活四者充分交融的宜人社区。

建筑设计

建筑采用新古典法式风格，用一种多元化的思考方式，将怀古的浪漫情怀与现代人对生活的需求相结合，兼容华贵典雅与时尚现代。设计一方面尊重和保留新材质和色彩的自然风格，摒弃过于复杂的肌理和装饰，简化线条，保持现代而简洁的审美倾向；同时，通过准确的比例调整和精致的细节设计很强烈地感受传统的历史痕迹与浑厚的文化底蕴，精致、端庄、对称。建筑立面处理细腻典雅，以体现自然、亲和的"家"的气质与形态。住宅立面采用新古典与现代结合的风格，与周围环境相协调，并丰富住宅建筑造型与户外空间效果。

户型设计

项目住宅产品覆盖了90~230 m²住宅单体。设计适应现代生活方式，强调大厅的活动空间，主卧室与起居室全部朝南，采用"十字"型空间，体现尊贵感。豪华户型充分考虑其端头的景观优势，设置有东西向的家庭室。同时，每套户型安排合理的半室外阳台空间，为住户实现空间的扩展提供一种灵活的解决方案。户型平面强调明厅、明厨、明卫及自然采光与通风的设计原则，厅房方正实用，统一设计室外空调机位、冷凝水管、厨房烟道、热水器排气口等，每户皆有工作阳台，主阳台进深不少于1.8 m，配有晒衣设施。

景观设计

景观设计整体采用法式古典园林风格。在主要轴线及出入口处采用庄重大气的景观处理手法，配以规则的水景、法式雕塑及修剪齐整的植栽，烘托出浓重的异域风情和高雅的格调。四个组团内部的景观处理则在法式风情的基础上巧妙地融入了中式园林的造景手法，兼顾了观赏与实用性的需求，营造出具有一定私密性的居家格调景观。整个园林以浅色系为基调，配以几十种植物搭配，给人以舒适宁静的感觉。绿化系统与道路系统设计相结合，将绿化引进各单体住宅。同时注意绿化节点的处理，在空间焦点与视觉转折处强化设计力度，注重草坪缓坡与休憩广场、流水、凉亭、雕塑等丰富空间对景的变化。

MASTER AND MASTERPIECE | 名家名盘

Profile

The project is located in the core of the development zone supported by the Nantong Municipal Government. It is 10 km away from the Nnatong downtown, with Zhenxing East road in its south, Xinjing Road in its east, and Xinkai River in its west. The project is next to the famous Wolf Mountain Scenic Spot and the Xinghu 101 Business District. It is designed as the French-style residential area in middle and high end for all ages, consisting of the high-rise residences, supporting retails and childcare facilities as well as convenient traffic and completed supporting facilities.

Planning Layout

The general layout takes the classic French-style axial symmetry as the core with the application of modern elements and Chinese-style garden landscape construction modes based on the classical architectural forms and the exotic landscape style. The site is divided into 4 zones of community groups, and the position of the architectures in each zone conforms to the topography, landscape and the change of direction, creating the interesting and charming spatial forms. Besides, the landscape is optimized to enrich the spatial levels, build a natural inhabitant environment for the community which is featured with space, architectures, landscape and living.

Architectural Design

The architectures with neo-classical French style and integrate the romantic meditation on the past with the modern life demand, which is luxurious, elegant and modern, fashion. The design maintains the natural style of the new materials and colors, and turns down the excessively complex texture and decoration to retain the aesthetic tendency of modern and conciseness with simplified lines. It also strongly expresses the traditional historical traces and deep cultural heritages by the precise ratio adjustment and the exquisite detail design in the exquisite, elegant and symmetrical way. The facades are delicate and elegant to present the quality and morphology of a natural and friendly "home". Their style is an integration of neo classics and modern, suitable with the surrounding environment, enriching the building shapes and the outdoor spatial effect.

House Layout Design

The project covers the residential units of 90-230 m², with the design of modern living way and the highlight of the activity space in the hall. The master bedrooms and living rooms are designed to be cross-shape space towards south, representing the sense of dignity. The luxurious houses fully take the advantage of the landscape in the end with the family room towards east and west. Semi-outdoor balcony space is arranged reasonably for each house as a flexible solution to the space extension. The plane of the houses emphasized the lighting in halls, kitchens and toilets, the natural lighting and the ventilation. The halls are square-shape and practical and the outdoor air conditioner's positions, the condensate pipes, the kitchen flues and the water heater exhaust ports etc are under unified arrangement. Every house owns a work balcony and the main balcony of 1.8m deep with clothes air-drying facilities.

Landscape Design

The landscape design adopts the French-style neo-classical garden style. The elegant and graceful landscape in the main axis and the entrances, along with the ordered waterscape, French-style sculptures and the manicured plantings, is highlighting a deep exotic and elegant style. The landscape in the four groups has adopted the Chinese gardening way on the basis of French style, to satisfy the demand of sightseeing and practicability, building the landscape with home style and privacy. The whole garden takes the light colors as the basis, collocating with tens of kinds of plants to provide a sense of comfort and peace. Greening system and road system work together to bring the greening into the unit houses. The design in space focus and visual turning points are strengthened and the attention is paid to the view changes in lawn gentle slopes, recreational squares, running waters, arbors and sculptures etc.

PROFOUND, AMBIENT, ELEGANT AND EXQUISITE | Taicang Green Land Town

沉稳 大气 典雅 精致 —— 太仓绿地新城

项目地点：中国江苏省太仓市
开 发 商：绿地集团
建筑设计：水石国际
项目规模：142 000 m²

Location: Taicang, Jiangsu, China
Developer: Green Land Group
Architectural Design: W&R Group
Project Area: 142,000 m²

项目概况

太仓绿地新城位于江苏省太仓市，处于城市的中央位置，周边配套齐全，地理位置优越。主要产品包括高层住宅、联排别墅及商业区等，属于绿地集团先期开发的住宅区的集中式社区商业，项目规模为14.2万m²。

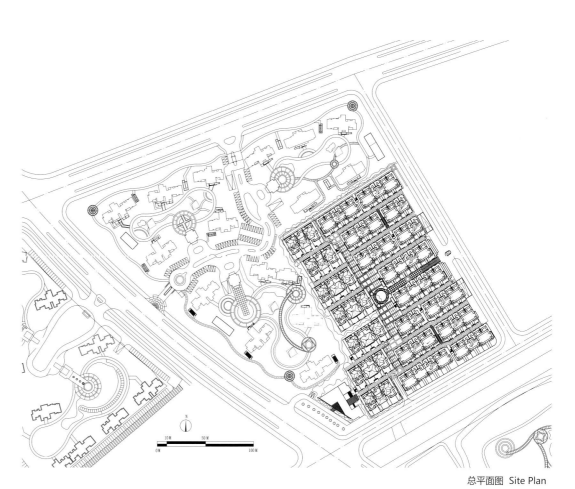

总平面图 Site Plan

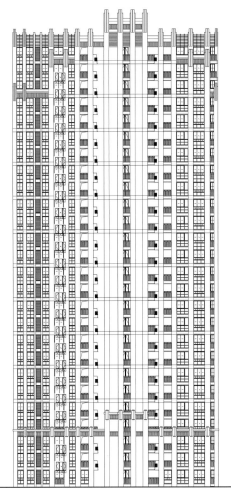
立面图 1 Elevation 1

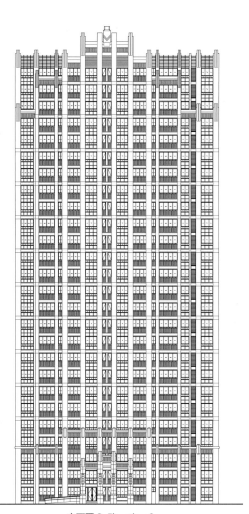
立面图 2 Elevation 2

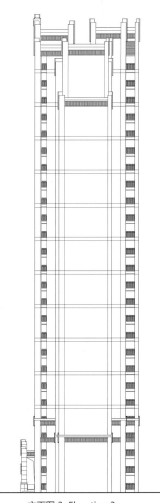
立面图 3 Elevation 3

规划布局

本案在规划布局上将零售区沿主干路设置，设三个入口与城市相接，主次明确标志鲜明；街区内部形成环状步行系统，流线清晰同时争取商业面最大化；四个中庭都有视觉标志物，成为各具特色的商业兴奋点；设置方便快捷扶梯电梯，实现双首层概念，强力拉动商业人流。

建筑设计

建筑立面采用石材、金属、玻璃材质，配以鲜艳的涂料，形成现代明快的外观特征。大卖场在地块深处临河布置，吸引目的性消费人群，与零售区之间设置空中连廊，拉动零售商业人气，临河立面采用彩色面砖结合巨幅广告灯箱，美化城市形象，突出建筑特征。商业街具有加建采光屋顶成为半室外空间的可能性；沿街设置LED显示屏提升商业气氛。

MASTER AND MASTERPIECE | 名家名盘

Profile

Taicang Green Land Town is located right in the center of Taicang City of Jiangsu Province, enjoying complete ambient infrastructure and superior geographic location. The main products include high-rise residence, townhouses and commercial area, etc, which belong to the centralized community type developed by Green Land Group in the early stage. The project covers an area of 142,000 m².

Planning and Layout

This project puts retail area along the main stem and three entrances connected with downtown, in clear and apparent order. Inside the block, it takes on ring-shape pedestrian system with clear streamlines for maximized commercial face. Each atrium possesses its own visual mark to become distinctive commercial interest. Convenient and rapid escalators are equipped to realize the concept of double-first floor for the drive of commercial stream.

Architectural Design

The elevations of architectures adopt stones, metals and glass with bright-color finishing to form a modern and lively appearance. The shopping mall is placed deep underground and next to the river, connected with retail area through an air bridge to attract potential consumers and drive commercial atmosphere. The riverfront elevation is equipped with colorful bricks and super large advertising light box, which helps beautify the urban image and highlight architectural features. The commercial street possesses the ability to wear lighting rooftop and become half outdoor while along-street LED display screen improves commercial atmosphere.

INTERVIEW | 专访

专一才能创造独特价值
—— 访安道（香港）景观与建筑设计有限公司
设计运营经理 宋晟

■ **人物简介**

宋晟
安道（香港）景观与建筑设计有限公司
设计运营经理
景观建筑研究中心经理
http://weibo.com/cheneysong

■ **公司简介**

A&I（安道国际）创立于2001年，是当代中国最具竞争力和创造力的景观建筑设计公司之一。十多年来，安道国际的设计作品跨越了品类、尺度和区域的限制，不论是精致氛围的度假社区、酒店、公园广场、城市社区，亦或规模宏大的商业综合体、科技办公及产业园、城市片区规划。A&I（安道国际）作为美国景观设计协会、美国建筑师协会和美国绿色组织协会的会员，始终与国际相关行业组织和国内外同行及高等院校保持着密切的互动关系。安道设计的魅力在于不断为每一位客户创造高品质的环境体验，为每一个作品注入新鲜的理念和创意，让设计服务能够超越客户的期望，实现商业价值与自然环境的和谐平衡。

■ **研究中心简介**

安道国际景观建筑研究中心成立于2010年，工作地点设于杭州，是国内为数不多的由设计公司创建的学术研究机构之一。安道国际景观建筑研究中心的工作旨在加强建筑、规划、景观三个人居环境学科领域的互动，在一定程度上推动着中国当代社会对景观建筑的关注。研究中心正在通过高等院校、学术媒体、行业协会等多个维度开展课题合作，并与美国景观建筑师协会（ASLA）、美国建筑师协会（AIA）、中国建筑学会、中央美术学院、北京林业大学、南京林业大学、《时代建筑》、《建筑细部》等多家单位保持着良好的工作往来。

《新楼盘》：请谈谈您对于景观设计的主张与思考，其中有哪些原则、元素、问题是您在设计实践中尤其关注的？

宋晟：我们的作品，皆源于对自然的种种体悟，追求建筑与环境的融合，空间与场所气质的统一。"景观"的本意是想表达人与场所之间的紧密关系，是一种与建筑物所协同的人造环境的表达。我们一直推行"大景观"的设计理念，以景观的视角理解和协调城市的发展，以景观作为载体介入城市结构，使其成为组织城市空间形态的重要手段。此外"建筑景观一体化"也是我们所一直坚持的，以"跨界"将景观从过去的从属地位上升到主导地位，以此引领未来城市发展的方向，在大景观的背景下规划城市与建筑。

《新楼盘》：能否结合您的项目经历，谈谈您认为风景设计师能为新城发展和旧城复兴做些什么？如何看待和衡量户外景观和公共空间所创造的价值？

宋晟：刚刚提到"大景观"的概念，我们所提倡的是以景观的角度来思考城市问题，以生态策略作为解决问题的切入点，无论是新城的发展还是旧城的复兴，都要以景观为载体，作为一种基础设施入到城市结构中去。安道执笔的东莞佛灵湖规划项目就是一例与自然相融的规划设计，将基地的自然特质理解为一种景观框架，在景观框架之上进行建筑的规划，设计中建筑不再突出而成为景观延续的公共空间。我们以景观的视角解决了新老城关系、城市空间扩张等一系列突出问题，创造出自然生动的城市形象。

在"大景观"的视角下户外景观和公共空间尤为重要。因为它们是城市的基底，建筑是填充在这些公共空间当中的附属物，在城市中的地位是主导的。城市户外空间和公共空间所创造的价值也是无与伦比的，它为人们提供了"呼吸"的场所，是符合人性的空间场所。

《新楼盘》：您如何考虑景观设计项目实现后的长期维护问题？针对这一问题，您有什么样的策略与经验？

宋晟：如果说一个建筑的完成是它生命走向倒计时的标志，那么景观的建成是它新生命的开始，景观会不断的成长而达到更加完美的效果，那么如何实现后期的完美景观效果，我认为植物材料的运用很重要，植物作为一种有生命的设计素材，会随着时间的推移产生时刻变化而又不断成长的景观效果，乡土植物的运用还会大大的降低养护的成本。此外，场所的文化内涵要特别强调，它可以延续和增强场所的生命与活力，最为一种隐性的要素，其作用不可小视，景观的特征通过独特的精神内涵得以淋漓尽致地体现，从而形成一种经久不衰的独特魅力与空间体验。

《新楼盘》：安道多年从业景观设计的基本理念就是以景观的思维做建筑，以建筑的视角看景观，以此谋求"跨界"的设计突破。能否谈谈为什么景观设计与建筑设计相结合的道路对于你们来说如此重要？

宋晟：安道从成立之初就把"建筑与环境景观的和谐统一"作为我们的设计理念之一。好的设计应该既不为突出景观，也不为彰显建筑，而是加强建筑、景观与周边的环境，与人的对话。当你模糊了建筑与景观的界限，你就能用一个更广阔的视角去关注我们的环境以及我们当下的生活，你会去思考人们真正的需求，这也是为什么我们在每个项目设计之前，就投入大量的时间对基地以及周边环境进行非常深入的调查与研究，有了这样一个深厚的基础，最终设计出的作品也能与城市的整体状况更加协调。相反，过分强调建筑与景观的独立性，往往会影响整个项目的品质，因此，我们积极谋求"跨界"的设计突破，把规划、建筑、景观三个专业的人员组织在一起，在建筑设计过程中，以景观规划作为指导，树立整体设计的观念，在景观设计过程中则以建筑设计的理性作为依托，这样我们的设计才会朝着健康的方向发展，才会永远走在时代的前列。

《新楼盘》：贵公司在高品质社区方面有很多优秀的作品，包括最近"千岛湖-桃花源"，荣获"第七届中国人居典范-最佳建筑设计方案金奖"。请您谈谈你们在不同类型的社区设计方面的理念和经验。

宋晟：安道特别提倡"人居之本"，也就是塑造真正适合城市人居住的环境。我们的每个社区设计，都力图探索"自然"与"家"的和谐秩序与内在联系，希望通过多变的设计手法，寻找出与每个特定空间相契合的表现方式，创造高品质的空间体验。度假社区千岛湖桃花源代表着安道在这方面的探索，设计通过营造广阔水域湖畔栖居的度假式休闲水岸生活方式，拉近人们与"自然事物"的距离，提供人与自然的思辨场地，从而让人们得到一种精神上的放松与自由，这样的探索我们还将一直持续下去。

《新楼盘》：目前，很多西方的设计师活跃在中国，西方的各种设计理念与设计手法对国内的影响很大，那么我们如何在这种情况下通过设计满足人们对中国本土文化、精神联系的追求？

宋晟：当代中国的景观理论大多引自西方，而我们的景观教育体系也存在着诸多混乱的现象，这样就造成了设计师具有"先天教育不均衡"的通病，对于传统文化与西方的思想没有一个清晰的思辨，于是因刻意追求现代气息致使设计类似或乏味，缺少文化根基的案例也屡见不鲜。好的设计作品一定是与当地的人文环境非常契合的，我认为我们可以在这几个方面下功夫：1.跳出"习惯'的思维，用异国人的视角看待和保护自己的文化，这样你会迸发出很多的设计灵感，会发现自己拥有很富足的文化。2.挖掘不同地域人的精神深层次的需求。3.在此基础上，我们要培养自身取其"形"，承其"意"，借其"神"，蕴其"情"的能力。通过最适当的形体，烘托出意境，体现出神韵，传递我们的情感。

《新楼盘》：作为诸多业界学会、协会的成员，贵公司频频出席各种行业内的大型学术活动。那么通过这些活动你们有何收获，由此可以看出国内外的景观设计实践各自有何特点？

宋晟：行业内的大型学术活动为我们同行之间提供了一个非常好的交流平台，既有助于增进彼此的了解，也大大拓宽了设计师的视野，通过多次学术交流可以看出国内外的景观设计实践还是存在一定差距的。中国式的城市景观正面临着三方面的挑战：可持续发展、文化身份、精神信仰的缺失等等。近年来中国城市的建筑热潮很多是人们的观念和权力在土地上的投射，中国的当代景观物化了人们脑海中的价值观和审美观，也严重背离了中国人的传统人文理念和乡土意识，而国外的设计师则对城市空间有更深刻的认识，他们重视生态绿化建设，注重环境可持续发展，对能源和材料的有效利用、对自然的治愈能力有很高的见解，这正是国内设计师应该积极努力的方向。

《新楼盘》：贵公司在去年刚刚进行了声势浩大的校园招聘活动，你们对新入职的员工有何期待？在员工培训和团队建设上又有哪些举措？

宋晟：公司非常重视对新入职员工的培养，我们有一套非常完善的人才培训体系，对于新入职的员工，我们精心安排了一系列的培训内容，包括：公司历史及企业文化培训、经典项目的案例介绍、不同的专业技能培训、团队沟通能力培训、经典项目实地参观等。期望通过我们的引导和帮助，让新员工能更快的适应新环境，尽快融入团队。为了拓宽团队视野，公司每年都会组织设计师外出学习，或有计划的安排去世界各地参观旅游。我们希望员工在认同公司企业文化的前提下坚持自己的个性，保持对设计的热情。

《新楼盘》：年轻的设计师在工作过程中，难免会遇到需要不断修改设计方案的情况。能不能结合您本人的经验谈谈如何才能在设计中获得成就感以保持设计的热情？

宋晟：景观设计是一个综合性很强的专业，在理性的思维指导下，也可以表现出时尚的跃动。你必须时刻保持一颗敏锐的心，不断地了解现在人们的生活方式及心理需求。所以在平时我会关注各种类型的国内外设计杂志、创意网站以及一些优秀的艺术作品展等等，从而不断提高自己在艺术及美学方面的修养以保证设计作品的生命力。

《新楼盘》：作为一家提供全面设计服务的专业化设计机构，贵公司刚刚走过了第一个十年，在国内完成上百项大型设计工程项目，在大型政府项目的国际招标中也多次获奖中标，并获得众多荣誉和奖项。那么您对公司的发展有何期望？

宋晟：作为有着高度社会责任感的公司，我们要做的不仅仅是区域性设计机构，更要做中国乃至世界具有强烈社会责任感的设计机构。面对国内景观行业充斥着巨大的泡沫，我们期待除了解决与景观设计相关的问题外，能以充满社会责任感的心感染这个行业，让与我们合作的伙伴们一起为生态的可持续做出贡献。

INTERVIEW | 专访

Only Concentration Can Create Unique Value
—— Interview with Song Cheng (Design Operation Manager of A&I International)

Profile:

About Song Cheng
Design Operation Manager of A&I International
Manager of Landscape Architecture Research Center

http://weibo.com/cheneysong

About A&I International:

Founded in 2011, A&I International is among the most competitive and creative landscape companies in China. For more than a decade, A&I's broke the restrictions on category, size and region, presenting great works covering resorts, hotels, parks, urban neighborhoods, large-scale commercial complex, sci-tech parks and urban planning. As the member of ASLA (American Society of Landscape Architects), AIA (American Institute of Architects) and USGBC (U.S Green Building Council), A&I has always kept a close contact with the international landscape and architectural organizations and schools. The power of A&I is to provide clients with high-quality environment experience, create every work with new concepts and ideas, make the design to serve the clients more than their expectation, and realize the balance between the commercial value and natural environment.

About Landscape Architecture Research Center:

Landscape Architecture Research Center of A&I International was set in 2010 with its office in Hangzhou. It is one of the few institutions for academic research which are established by design companies. The goal of this center is to strengthen the bond between the subjects concerning living environment—architecture, planning and landscape. It tries to draw public attention to landscape architecture. The research center is now promoting the cooperation with universities, academic medias and professional associations. In addition, it keeps good working relations with ASLA, AIA, Architectural Society of China, China Central Academy of Fine Arts, Beijing Forestry University, Nanjing Forestry University, Time + Architecture, DETAIL and so on.

New House: For landscape design, would you like to share some opinions and thinkings with us? And what're the principles, elements and problems that you concern more in your practices?

Song: All of our works are inspired by the thinking to nature. And we always pursue the harmony between architecture and environment as well as between space and atmosphere. The aim of "landscape" is to show the close relationship between human and the environment. We always insist on the design idea of "Macro Landscape" to interpret and coordinate the development of cities with the landscape. Landscape will be the carrier to urban structure and an important element to organize the urban space. In addition, "integration of architecture and landscape" is also what we insist. Landscape to architecture, from dependence to independence, leads the way to future urban development.

New House: As we all know that you have rich experience in landscape practice, and what to you think the landscape designers can do in the new town development and old town renovation? How to measure the value of outdoor landscape and public space?

Song: Above I've mentioned the concept of "Macro Landscape", and we encourage the way to solve urban problems with landscape design and ecological strategies. No matter for the new town development or for the old town renovation, landscape will be an important element to urban structure. Take A&I's project — Dongguan Foling Lake Planing for example, it combines with the natural environment successfully. The characteristics of the site are taken as the landscape frame and the architectural planning is made accordingly. Architecture will be the extension of landscape space but not the only focus. Thus we have coordinated the relation between new and old towns, solved the problems in urban expansion, and gave a particular city impression.

According to the concept of "Macro Landscape", outdoor landscape and public space are quite important. They are the basic elements of a city, and all the buildings are designed to fill these public spaces. And the values they create are incomparable because they are to humane spaces for "breathing".

New House: How do you consider the long-term maintenance after the realization of a landscape design? Would you like to share some strategy or experience?

Song: If we say the completion of a building is a sign means the decay of its life, while the completion of a landscape is the beginning of a new life. Landscape will continue to grow and achieve perfect condition as time goes by, then how to get the best landscape effect in later stage, it is very important to select plants in my eyes. As living design materials, plants would keep growing day by day to present better landscape and the use of native plants will greatly reduce the cost of conservation. In addition, attention should be paid to the cultural connotation of a site, for it can extend and enhance the life of the site, as the most hidden element, it plays a not-so-trivial role. Unique spiritual connotation may reflects the landscape characteristics incisively and vividly to form a long-standing certain charm and spatial experience.

New House: In order to seek a "cross-border" design breakthrough, designing landscape from architecture and seeing architecture from landscape is the fundamental philosophy of A&I in these years. Can you talk about why it is so important to you to combine landscape design and architectural design?

Song: From the beginning, A&I regarded "architecture and environment landscape should be in harmony and unity" as one of the design philosophies. Good design should be neither prominent landscape nor highlight building, but to strengthen the dialogue between them and the surrounding environment and people. When you blur the boundaries of architecture and landscape, you will be able to focus on our environment and our life in the present in a broader perspective, and to think about people's real needs. That's why we make deep and further survey and research on the site and surrounding environment before designing a project. Only by this solid foundation, can the final design get well with the overall condition. On the contrary, too much emphasis on the independence of architecture and landscape often affect the quality of the entire project, therefore, we are actively seeking "cross-border" design breakthrough, gathering the professionals in three fields, planning, architecture and landscape. In terms of architectural design, we take landscape planning as a guide to establish the overall design concept. And we reply on the ration of architectural design when designing the landscape. Only by doing so can we develop in the right direction and always walk in the forefront of the times.

New House: Your company has presented many excellent works on high-quality community, for example, "Qiandao Lake- Shangrila" which has recently won the golden prize of "Seventh Chinese Habitant Paragon-The Best Architectural Design Proposal". Could you share with us your design concept and experience of different types of communities?

Song: A&I is highly calling for "the base of inhabitance" —constructing the environments exactly suitable for urban inhabitants. We endeavor to exploit the harmony and interior relationship between nature and house in every community design, try to work out the appropriate modes fitting into specific space by multivariate design practices, so as to create a high-quality spatial experience. The project features our exploration on this aspect. A resort-style waterfront lifestyle with vast water and lakeside dwelling is designed to bring people closer to nature, and offer the meditation site for their spirit relaxation and freedom. We will keep working on this kind of exploration.

New House: Currently many western designers join in the China market, their design concepts and practices have brought a great influence in China. In this case how to satisfy the demand for Chinese native culture and spirit connection through design?

Song: Most of the contemporary landscape theories in China are learned from the western, as well as the chaotic in our landscape educational system, our designers are receiving the unbalanced congenital education. Lack of clear speculation about the traditional culture and western minds, they overemphasize the modern character and work out similar and boring designs without culture fundament, which is nothing new. An excellent design proposal has to correspond to the local humane environment; I think we could work in the following ways: 1. Standing out of the "custom" thinking pattern, observing and protecting out culture in the exotic shoes, which will lead to great design inspiration and the awareness of abundant culture; 2. Exploiting the further sprit demand from different districts; 3. Learning more to be capable of expressing the idea and emotion through the most proper shapes.

New House: As member of many professional academies and associations, your company has attended a lot of famous academic events and activities. What do you get from these events and activities, and what are the characteristics of the landscape practices by Chinese designers and foreign designers?

Song: The large academic activities within the industry provides us a very good communication platform, which helps promote each other's understanding, also greatly widened the designer's horizon; through so many academic communications we can see that the landscape design practice at home and abroad still exists a certain gap. The Chinese urban landscape is facing three challenges: sustainable development, cultural identity, the lack of spiritual beliefs and so on. In recent years, Chinese urban construction boom is the projection of people's idea and power on the land; Chinese contemporary landscape materializes people's mind, values and aesthetic view, also seriously deviates from the Chinese traditional humanistic ideas and local consciousness. However, foreign designers have a more profound understanding of the urban space. They pay attention to the ecological green construction, the sustainable development of environment, the effective use of energy and materials, and have a great opinion of the nature healing ability, which is the way domestic designers should try to follow.

New House: Your company has just held a big campus recruitment in the last year, and what is your expectation for the new employees? What measures will you adopt in the staff training and team construction?

Song: Our company attaches great importance to the new staff training, we have a very complete training system; for the new appointed staff, we arrange a series of training courses, including the company history and enterprise culture training, classical project case introduction, different professional skills training, team communication skills training, classical project field visit on-site, etc. Through our guidance and help, we help the new employees adapt the new environment and integrate into the team as soon as possible. In order to broaden the horizon of the team, we organize the learning exchange for the designers, or make a visit or tour to all over the world. We hope that the staff insist on their own personality and keep the enthusiasm for design under the premise of identifying with the enterprise culture.

New House: Young architects usually encounter some cases which they have to modify their designs a few times. Would you please talk about how to gain sense of achievement and maintain passion for design according to your own work experience?

Song: Landscape design is a comprehensive career, which could express fashionable motions under logic thinking. You have to keep a sensitive heart at all times and try to find out people's way of living and their psychological needs. So usually I'm used to following all kinds of foreign design magazines, creative website and some fine arts show in order to improve my accomplishments in art and aesthetics to help maintain the vitality of my works.

New House: As a professional design company providing comprehensive design service, you have achieved your first 10-year milestone and finished nearly a hundred pieces of large domestic projects. And you also won several international bids for large governmental projects which returned you with honor and prize. May I ask what your expectations are for your company?

Song: As a company with high sense of social responsibility, our goal does not rest on regional projects, but national or even international. Faced with the strong foam in domestic landscape design circle, we expect not only to solve problems related with our professional, but also to affect the entire industry with our sense of social responsibility as well. We are looking forward to make contributions for the sustainable development of ecology with our counterparts.

3D ECOLOGICAL SPACE WITH TRANSPARENCY | Apartment Building
强调立体感与通透性的生态化空间—— 公寓大楼

项目地点：意大利卡利亚里墨西拿路
建筑设计：意大利Dante O. Benini建筑师事务所
建 筑 师：Dante O. Benini, Luca Gonzo
摄　　影：Beppe Raso

Location: via Messina, Cagliari, Italy
Architectural Design: Dante O. Benini & Partners Architects
Architects: Dante O. Benini, Luca Gonzo
Photography: Beppe Raso

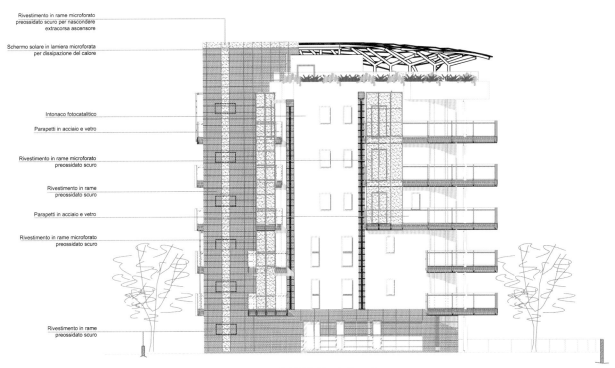

北立面图 North Elevation

NEW LANDSCAPE | 新景观

项目所在区域位于卡利亚里（撒丁岛）的博尔扎诺路与墨西拿路之间。该项目原是城市复苏计划的一部分，其建筑施工在进行到60%时一度被地产经营者叫停，而后Dante O. Benini建筑师事务所受命，项目得以完工。

公寓大楼依崖脚而建，面朝卡利亚里海湾，地理位置优越，美中不足的是大楼附近的空地上有一个当局留下的通风井，像是给地块添了一道伤疤似的。

项目后期尊重前期的框架，在原有基础上强调聚合和立体构成，以达到布鲁诺·赛维教授定义的四维空间——动态空间。每个立面既遵循原计划的形态，也展现出各自的特点。建筑师仔细分析风向和建筑朝向之后，采用了他们一贯青睐的建筑材料：穿孔金属板和网眼，打造可持续建筑，这些孔眼在保证室内可观景的同时也屏蔽了由外而内的视线。

公寓包括地下2层、地上6层和一个日光浴室……钢结构部分含钢化玻璃和预氧化铜；此外，还配有一些高效能设施，如冷凝式锅炉，太阳能面板和VRF（可变冷媒流量）空调，以确保新建筑的生态持久，尽量避免频繁的特别养护。

NEW LANDSCAPE | 新景观

Real estate operator, charged of urban recovery plan of the area between Bolzano and Messina roads in Cagliari (Sardinia) decides to stop a building construction still at 60% rough-built state in that area and charges DOBP to continue the whole area re-qualification.

The building is at the base of a cliff facing Cagliari bay, in a very privileged position but in front of an open air pit left by the municipality like a territorial injury.

It was decided for no modifications of what built, but to work on aggregation basis without exceeding in superfetations, and working on 3D composition, to achieve as per Professor Zevi definition the fourth dimension: "space dynamic" Each facade is different from any other, while maintaining the organic morphology at the project origin. A careful analysis of winds and the building orientation has advantaged the use of some basic favorite materials in DOBP projects: micro perforated metal sheets and meshes used to carry out a sustainable architecture from the energetic point of view, creating anyway a shield from outside view without preventing outside sight from inside.

The residential building has 2 underground levels, 6 floors and a solarium… Its steelwork structure is completed with treated glass and pre-oxidized copper; in addition high efficiency facilities such as condensing boiler, solar panels and VRF (Variable Refrigerant Flow) air-conditioning give rise to a new generation building assuring eco-sustainability, duration and very little extraordinary maintenance.

FLOATING MUSIC, DREAMY LIFE
| Shanghai Palaearctic Sheshan Villa Landscape Design

流动的音乐 梦想的生活 —— 上海古北佘山别墅景观设计

项目地点：中国上海市
景观设计：上海唯美景观工程设计有限公司
总用地面积：1 200 m²

Location: Shanghai, China
Landscape Design: Shanghai WEME landscape Engineering Co., Ltd
Land Area: 1, 200 m²

　　项目位于上海赵巷古北佘山，处于嘉松公路以西，沪青平高速以南，紧邻北竿山鸟类保护区，别墅建筑为新古典主义风格，坐北朝南，曲水相依，配套设施齐全。

　　设计灵感缘起"建筑是凝固的音乐"，从别墅建筑衍生灵感，以新古典主义音乐家斯特拉文斯基的代表作D大调小提琴协奏曲为设计母题，依照流畅变换的乐章结构来谋篇布局，赋予庭院不同空间以不同的主题。别墅花园以丰富室外活动功能为设计亮点，倾力打造"花园中的起居室"，营造自然、典雅富于文化气息的户外梦想家园。

流畅变换的华美乐章——动静相宜的空间组织

　　D大调小提琴协奏曲分四个不同主题的乐章。音乐以第一乐章欢快有力的音符开始，逐渐引出柔美律动的第二乐章咏叹调；然后在不断重复中逐渐上升到浪漫抒情的第三乐章，最后在轻松并富于生气的第四乐章中将整个乐曲推向高潮。设计师用通感和联觉把音乐和花园空间两种不同的媒体完美的结合，绿荫草地的入口开放空间，欧式水钵作为中心水景，突出欢快活泼的别墅入口空间意向。半私密半围合的木构架步道，柔美自然，犹如一唱三叹、盘旋上升的咏叹调。后园私密安全，动静相宜，集中了家庭聚会，休闲社交的主要功能，宛如一部高潮迭起的随想曲，气韵流淌，别具风韵。

花园中的起居室，自然中的家

　　庭院是别墅生活的另一半。设计师以花园起居室的建筑概念对庭院进行功能分区，往来交通

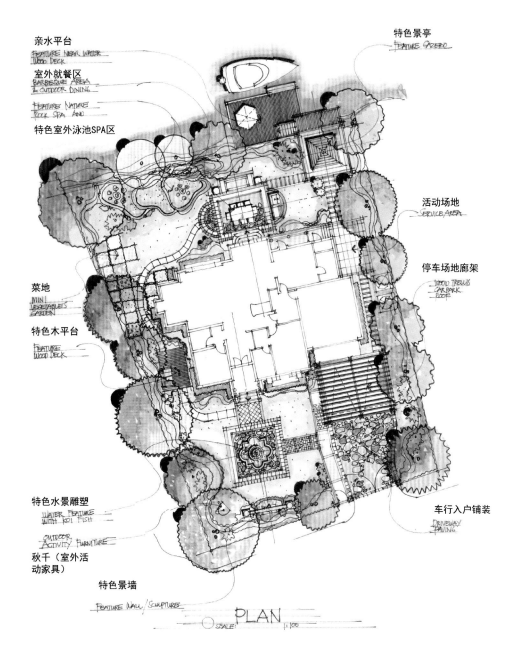

NEW LANDSCAPE | 新景观

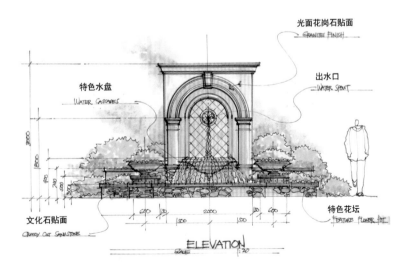

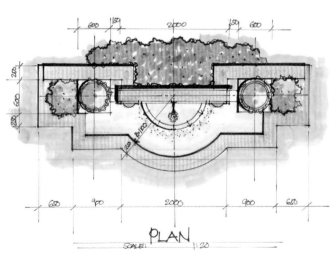

的前院好像门厅，一组景墙作为前门的对景，欢迎主人的归来。木廊架串联起前厅和后院，原木材质结合植物造景，浑然天成，亲切娴静，宛如优雅的小书房。后院集餐厅、会客室、运动场、菜园于一身，承担了业主日常活动的主要功能，聚餐烧烤、休憩眺望、微型高尔夫运动，游船码头，不同活动空间相互联系又各自独立，满足了业主家庭成员多层次的使用需求，打造了充满阳光、美好浪漫的梦想家园。

贯穿始终的优美和弦——壁炉和水景元素

协奏曲的另一个经典之处在于贯穿始终的优美和弦，设计师将这种音乐的连续性做到了别墅景观里。用新古典主义不可缺少的壁炉作为联系室内外景观元素的显性线索，以无处不在的精致水景作为隐性线索，穿插呼应，室内外的景观仿佛融为一体。人们居住其中，感觉不到自然在哪里终了，艺术从哪里开始。

别墅花园代表的不仅是身份地位的尊荣，更是一份生活方式的洒脱。本案设计追求在优美景观、优雅的音乐氛围的同时赋予花园更丰富的户外使用功能，让别墅生活在梦幻户外活动中更为自然与健康，在音乐与美景中更加富足与安逸。

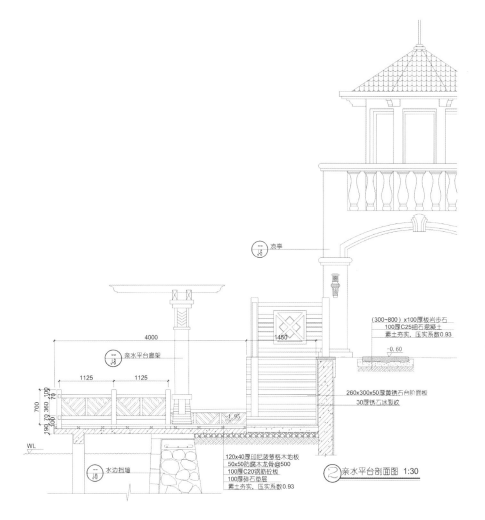

NEW LANDSCAPE | 新景观

The project is located in Palaearctic Sheshan at Zhaolane, Shanghai, on the western side of Jiasong Road, southern side of Huqingping Expressway, close to Beigan Shan bird sanctuary. The villa architecture faces to the south, with the hill and the lake nearby, with complete ancillary facility, and which is of a neoclassical style.

The designers derive inspiration from villa architecture, take new classical musician Stravinsky's masterpiece Violin Concerto in D as design motif, and the entire layout give different theme to different spaces of the courtyard according to fluent and converted movement composition. The design of villa garden highlights in abundant outdoor activities function, strive to create "the living room in the garden", build a dreamy home with nature, elegance, full of culture feeling.
Fluent and Converted Gorgeous Movement——Dynamic and Competitive Space Organization
Violin Concerto in D divided into 4 movements with different themes. The first movement started with cheerful and powerful musical note, the second movement gradually leaded to morbidezza rhythm aria, and then the third movement ascended to romantic lyric in

continuous repeat, lastly the relaxed and exuberant forth movement push the whole musical composition to climax. The designers combine music and garden space via synesthesia, the entrance of the shade grassland is a semi-closed space, western water bowel as its centre waterscape highlights cheerful and lively villa entrance space. Semi-illicit and semi-closed timber frame footpath is graceful and natural as one word, three sighs style, spiral rising aria. The private and secure backyard gathered the main function of family party and social intercourse dynamic and competitive, which like a piece of capriccio with one climax after another, flowing artistic concept and unique charm.

Living Room in Garden, Home in Nature

Courtyard is the other half of villa life. The designers considered the garden living room as the architectural concept to divide the space functionally; the forecourt for traffic circulation is like a hall, a group of feature wall as the opposite scenery of front gate to welcome the owner. The embellished frame connect antechamber and backyard, the material of log integrate in plant landscaping, like a sculpture made by the nature, amiable and demure, like a elegant small study. The backcourt with gathering hall, reception room, playground and vegetable garden in it, functioned as owner's daily activities site, BBQ dinner, rest for overlook, mini-golf, marina connected with each other but independent, satisfied family members' demand, became a sunshine, graceful and romantic dreamy house.

Throughout Graceful Chord——Fireplace and Waterscape Elements

Another classic of concerto is throughout graceful chord, the designers applied musical continuity into villa landscape. Using the fireplace which is necessary in neoclassicism as dominant clue, the fireplace connects interior or exterior landscape element, and using the ubiquitous delicate waterscape as recessive clue, via penetrating and corresponding with each other the interior and exterior landscape as if fused together. People can't feel where the ending of nature is and where the beginning of art is when living in this villa.

Villa garden not only stands for dignitary status, but also stands for free and easy life style. The design of this project give this garden plentiful outdoor using function when perusing beautiful landscape and elegant musical atmospheres, let villa life be more natural and health in fantasy outdoor activities, be more satisfied and comfortable in music and graceful scenery.

FEATURE | 专题

高层公寓

专题导语

随着城市化进程的不断推进，城市中心的可用地面积越来越少，催生了城市建筑必须由横向扩张模式转向竖向空间的发展模式。在这一背景下，发展高层建筑成为了城市发展的必经之路。高层公寓作为城市中心建筑的重要组成部分，在提高土地利用价值、满足城市居住需求以及优化城市建筑结构等方面发挥着重要的作用。如今的高层公寓已经从关注基本层面的设计内容转向更大范围的城市设计、更深层次的人居需求、更精细化的住宅设计以及更优美的社区生态景观设计等，其设计要求也逐渐体现出整体性、系统性、精细化的趋势。如何把握高层公寓的立面风格、体现景观特色、创新户型结构等也成为设计师普遍关注的问题。本期专题我们将与您一同去关注和探讨高层公寓中的诸多设计细节。

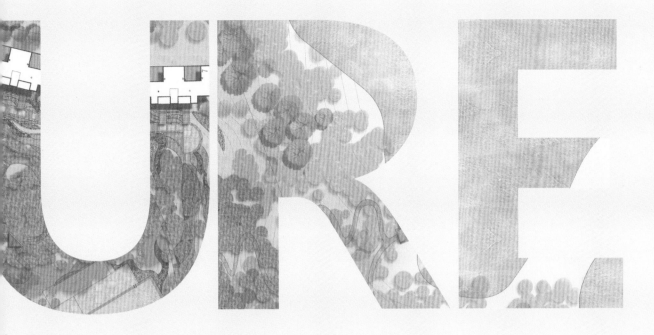

Introduction

With the continuous progress of the urbanization, the less available land area in the city center gave birth to a new mode of urban architecture, which focuses on vertical expansion rather than the original lateral spread. In this context, the development of high-rise building has become the only way of future urbanization. As an important part of the buildings in the city center, high-rise apartment plays an important role in enhancing land value, meeting the living needs of the citizens and optimizing the structure of urban architecture. Nowadays, high-rise apartment is no longer a building that just pays attention to basic level, but a fine residence in larger scale that could meet the deeper living needs and even an ecological community with beautiful landscape, reflecting a tendency towards integrity, systematicness and elaboration. How to grasp the facade style of high-rise apartment, reflect the landscape features and innovate the house structure has also become a common concern of the designers. We will be with you to explore the design details of high-rise apartment in this issue.

上海保利房地产开发有限公司
POLY SHANGHAI REAL ESTATE DEVELOPMENT CO., LTD.

企业介绍

上海保利房地产开发有限公司成立于2003年，注册资本金1亿元人民币，是保利房地产（集团）股份有限公司在上海业务的运作平台。公司下辖六个项目公司：上海建乔房地产有限公司、上海君兰置业有限公司、上海保利建锦房地产有限公司、上海城乾房地产开发有限公司、上海保利建霖房地产有限公司和上海保利建雍房地产有限公司。

产品与服务

保利地产坚持以商品住宅开发为主，适度发展持有经营性物业。公司目前在上海已开发六个住宅项目，分别是保利•十二橡树庄园、保利•香槟国际、保利•海上五月花、保利•西子湾、保利•叶上海和保利•林语溪。公司奉行"和者筑善"的企业价值观，将"和谐"提升至企业品牌战略的高度，致力于创造自然、建筑、人文交融的和谐人居生活。将"和谐"理念始终贯彻于企业的规划设计、开发建设和客户服务全过程，研创节能环保、自然舒适的产品，提升产品品质，通过亲情和院式服务营造良好的社区氛围，赢得了消费者的广泛喜爱。

Company Profile

Established in 2003 with a registered capital of 100 million yuan, Poly Shanghai Real Estate Development Co., Ltd. is a part of Poly Real Estate Group that based in Shanghai. It governs six companies: Shanghai Jianqiao Real Estate Co., Ltd., Shanghai Junlan Real Estate Co., Ltd., Shanghai Poly Jianjin Real Estate Co., Ltd., Shanghai Chengqian Real Estate Development Co., Ltd., Shanghai Poly Jianlin Real Estate Co., Ltd. and Shanghai Poly Jianyong Real Estate Co., Ltd.

Product and Service

Based on developing commercial residence, Poly Real Estate develops business property appropriately. At present, it has developed six residential projects: Poly Twelve Oaks Manor, Poly Champagne International, Poly May Flower, Poly Sizihwan, Poly SH-Culture Forest and Poly Ripple Impression. The company pursues corporate values "harmony makes perfect" and is committed to creating a harmonious living atmosphere that integrates nature, building and humanity. The "harmony" philosophy has always been consistent in the whole process from planning and design, development and construction to customer service. These works are energy saving and environment friendly. The favorable atmosphere and high-quality service they provide are quite popular among the consumers.

武汉万科房地产开发有限公司
WUHAN VANKE REAL ESTATE DEVELOPMENT CO., LTD.

企业介绍

武汉万科房地产开发有限公司成立于2001年，是万科集团在武汉设立的一个分公司。万科企业股份有限公司成立于1984年5月，1988年进入房地产领域并以其作为公司核心业务，并于当年在深圳证券交易所挂牌上市，是中国大陆首批公开上市的企业之一。迄今为止，万科已进入全国42个大中城市，并确定了以珠江三角洲、长江三角洲、环渤海湾、中西部区域四大城市经济圈的发展策略。凭借一贯的创新精神及专业开发优势，万科树立了中国知名住宅品牌，并为投资者带来了稳定增长的回报。

产品与服务

经过多年多潜心耕耘，至2012初，万科在武汉已连续开发四季花城、城市花园、西半岛、香港路8号、万科润园、万科魅力之城、金色家园、高尔夫城市花园、万科圆方、万科金域华府、万科城、红郡、金色城市、金域蓝湾、汉阳国际等多个项目，实现了武汉三镇布局。在为无数家庭提供优质环境的同时，武汉万科及其项目陆续获得了来自建设部、湖北省、武汉市等多个部门颁发的几十个奖项，并连续多年成为武汉市房地产十强企业，在武汉市民心目中树立了良好的形象。

Company Profile

Established in 2001, Wuhan Vanke Real Estate Development Co., Ltd. is a branch of Vanke Group. Vanke Group was established in May 1984, entered the real estate field and made it as the company's core business in 1988, and then was listed on the Shenzhen Stock Exchange at the same year. It is one of the first batch of mainland China publicly listed companies. So far, Vanke has entered the country's 42 large and medium-sized cities, and establish the four large Urban Economic Circles development strategy including Pearl River Delta, Yangtze River Delta, Bohai Bay and the Midwest Region. With consistent spirit of innovation and professional development advantages, Vanke has established a well-known residential brand in China and made a steady growth return for investors.

Product and Service

With years of concentrated work, Vanke has continuously developed Four-season Flower City, City Garden, West Peninsula, Hong Kong Road No.8, Vanke Rain Garden, Glamour City, Golden Home, Golf City Garden, Vanke Yuanfang, Vanke Businessmen Private Mansion, Vanke City, Stratford, Golden City, Paradiso and Hanyang International in Wuhan till early 2012, which realizes its layout in Wuhan. While providing a quality environment for countless families, Wuhan Vanke and its projects gradually won dozens of awards issued by a number of departments from the Ministry of Construction, Hubei Province and Wuhan City. It has been a top ten enterprise in this field for many years and established a good image in this city.

| 项目地点：中国湖北省武汉市 |
| 开 发 商：武汉市万科房地产有限公司 |
| 建筑设计：兰闽建筑师事务所 |

RANDOM SKIN AND DYNAMIC SPACE | Vanke Golden Homes, Wuhan

错跃的表皮结构 轻松动感的空间 —— 武汉万科金色家园

项目地点：中国湖北省武汉市
开 发 商：武汉市万科房地产有限公司
建筑设计：兰闽建筑师事务所
设计人员：兰闽、钟志怡、戎晓群、刘登举、顾波
施 工 图：武汉市建筑设计院

Location: Wuhan, Hubei, China
Developer: Wuhan Vanke Real Estate Co., Ltd.
Architectural Design: LM Architects
Designer: Lan Min, Zhong Zhiyi, Rong Xiaoqun, Liu Dengju, Gu Bo
Construction Plans: Wuhan Architectural Design Institute

项目概况

　　项目位于武汉市繁华的旧城区，地处武汉内环线内，被武汉大道分割成南北两部分。其中的北地块作为一期开发，承担着带动整个项目商业氛围的任务。距江汉路、武广商圈不到1km，周边配套非常齐全，交通体系完善、便捷，是万科第一个城市内环线以内的项目。

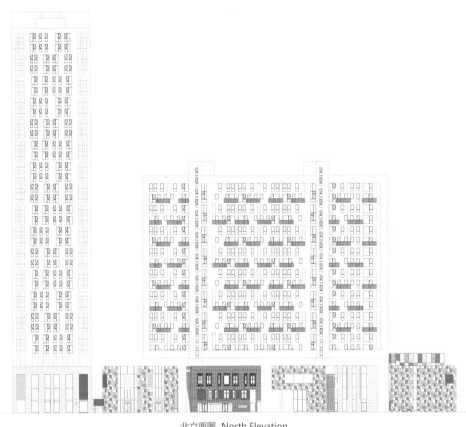

北立面图 North Elevation

南立面图 South Elevation

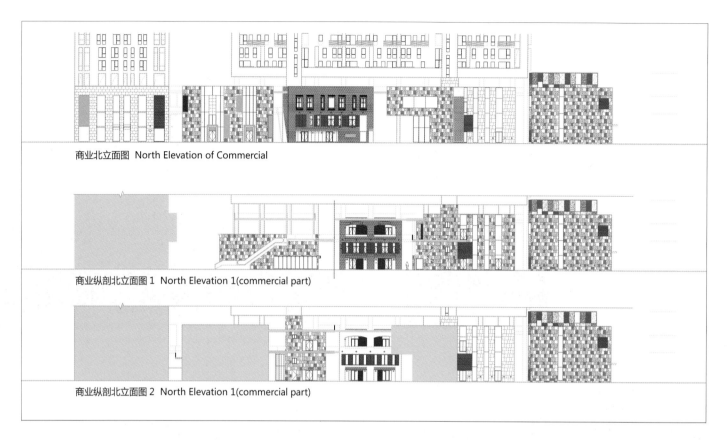

商业北立面图 North Elevation of Commercial

商业纵剖北立面图 1 North Elevation 1(commercial part)

商业纵剖北立面图 2 North Elevation 1(commercial part)

规划布局

项目沿袭万科城市住宅户型设计、城市泛会所、城市共生的精髓，以现代主义建筑构思和审美品位，追求建筑的质感，追求空间尺度的精细，在户型以及入户方式等方面进行了研究创新。项目规划的地面一、二层的商业设施均为小型店铺，它们被规划为九个独立的单元，通过立体街道将其连通，每个单元有不同的装饰风格。西侧的玻璃体是住宅入口，有观光电梯将住户由地面运送到商业顶部的架空层，再沿水平方向到达住宅电梯。

建筑设计

尽管项目规模不大，但在设计上却极为复杂。板式住宅的户型类似于马赛公寓的"错跃"结构，其南侧立面上错动的玻璃方块成为主要的形式单元，它集合了以下的重要功能：作为露台赠送；24m高的玻璃面屏蔽噪音；遮阳、晾衣。东侧塔楼的形式设计则以一种朴素轻松的方式融入一个混乱、破旧的城市。

户型设计

万科金色家园主力户型为2室1厅1卫67 m²、2室1厅1卫85 m²、2室1厅1卫87 m²、2室2厅1卫90 m²、3室2厅2卫106 m²。

FEATURE | 专题

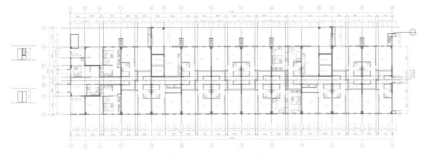

板楼 7、11、15、19 平面图
7th, 11th, 15th and 19th Floor Plan

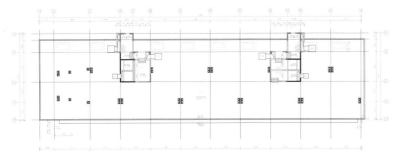

板楼顶层平面图 Top Floor Plan

Profile

Located in the old downtown area of Wuhan City, within the inner ring, the site is divided into two parts by Wuhan Avenue. The north part is for phase I, responsible for shaping the commercial atmosphere for the whole project. Near to Jianghan Road and Wuguang CBD within 1km, enjoying varied supporting facilities and perfect traffic system, it is the first Vanke project that locates within urban ring.

Planning and Layout

The project continues Vanke's tradition to purse aesthetic taste, architectural quality and delicate space. The design for housing types and hallways was researched and renovated. According to the planning, the first floor and second floor of the commercial facilities are used for small stores which are arranged in nine independent units. They are connected by three dimensional streets and passages. Each unit is decorated in unique style. The glass volume on the west side is the entrance to the residences. There is sightseeing lift transferring residents to the elevated floor on top of the commercial podium, and then there are residential lifts waiting for them.

Architectural Design

Though a moderate size, the design of the project is extremely complicated. The housing types of slab-type houses adopt a random structure that's similar to Marseille apartment. The interlaced glass cube on the south facade is the main units which can be used as terrace and balcony. And the 24 m high glass facade can well insulate the noises and protect the building from overheating. While the tower in the east looks elegant and relaxing in this busy city.

Housing Type Design

The main housing types of Golden Homes are 67 m^2 two rooms, 85 m^2 two rooms, 87 m^2 two rooms, 90 m^2 two rooms and 106 m^2 three rooms.

DYNAMIC MODERN SAIL-SHAPED BUILIDING | Jakarta Luxury Apartments, Indonesia
极富动感与现代元素的帆船状建筑——印度尼西亚雅加达豪华公寓

项目地点：印度尼西亚雅加达
客　　户：Badan Kerjasama Mutiara Buana
建筑设计：阿特金斯
占地面积：376 550 m²

Location: Jakartta, Indonesia
Client: Badan Kerjasama Mutiara Buana
Architectural Design: ATKINS
Occupancy Area: 376,550 m²

项目概况

该项目涵盖五星级酒店，豪华公寓以及酒店式公寓建筑，并配备了水上乐园，地下停车场等设施。设计师旨在将项目建成顶级公寓，并将其打造成雅加达滨海区的地标建筑。

建筑设计

在建筑设计方面，设计师运用航海这一设计理念，将建筑定位为海上航行的指南针。公寓建筑群代表游艇根据指南针的指示行至远方。公寓外立面帆船形状的建筑侧鳍突出远航的特征。设计理念简单而富有动感，该项目目前已成为珍珠海滩的地标建筑。

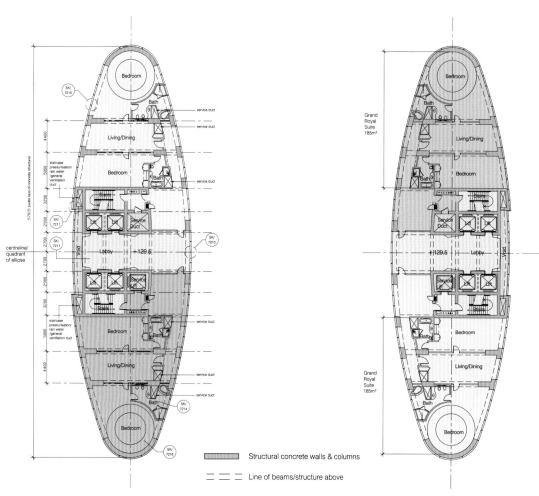

总体规划图 Master Plan

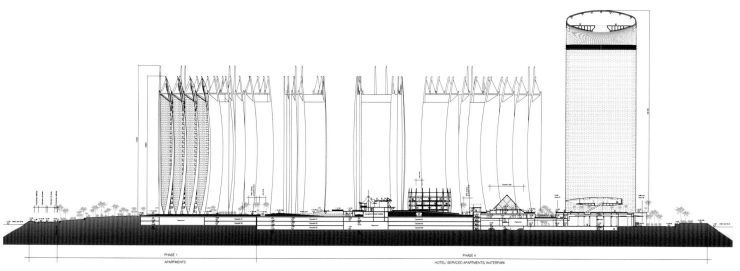

总剖面图 BB Master Site Section BB

FEATURE | 专题

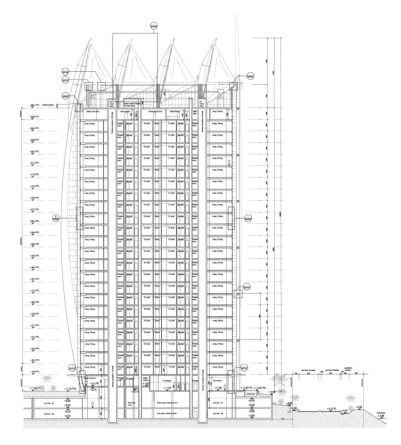

剖面图 B-B Section B-B

Profile

The project comprises a 5-star Tower Hotel, 10 luxury category apartment towers, 2 serviced apartments, water park and basement parking facilities. The project brief was to target the top level of the high class apartment market and architecturally the development to be a landmark on the Jakarta waterfront.

Architectural Design

The concept was to use a nautical theme and orientate the buildings on the cardinal points of the compass thus creating the best possible views of the waterfront for each phase of development. The cluster of apartment buildings represents elegant yachts sailing for distant locations on the cardinal points of the compass. The sailshaped architectural fin treatment on the apartment facade also emphasises the nautical character for each apartment cluster. The concept is simple but dynamic and the development is now a landmark on the Pantai Mutiara canal estate when viewed from land, air and sea.

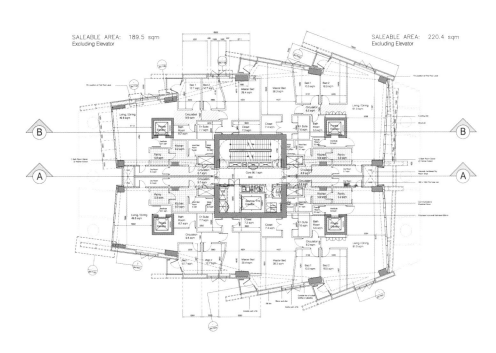

典型一层平面图 Typical First Floor Plan

PASSIVE RESIDENCE WITH DIVERSIFIED ELEVATIONS
Air Apartments
多元化立面构造下的被动式住宅——空中公寓

项目地点：澳大利亚昆士兰
建筑设计：Ian Moore建筑师事务所
占地面积：21 170 m²
总建筑面积：22 603 m²
摄　　影：Rocket Mattler

Location: Queensland, Australia
Architects: Ian Moore Architects
Site Area: 21,170 m²
Gross Floor Area: 22,603 m²
Photography: Rocket Mattler

项目概况

　　公寓位于区域的东部边缘上，黄金海岸的背面，冲浪天堂的南面，可以独享海洋和沙滩的壮丽景色。同时，黄金海岸西面的景色以及大分水岭可以尽收眼底。

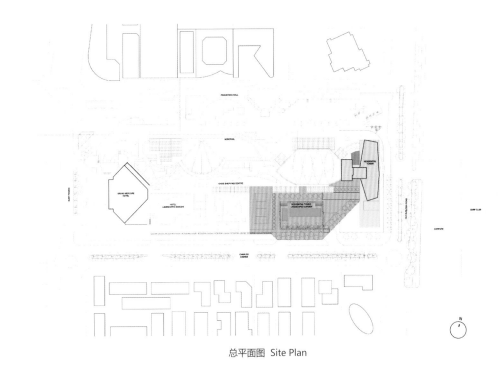

总平面图 Site Plan

WEST ELVATION - SECTION　　　　　　　EAST ELEVATION - OLD BURLEIGH ROAD

FEATURE | 专题

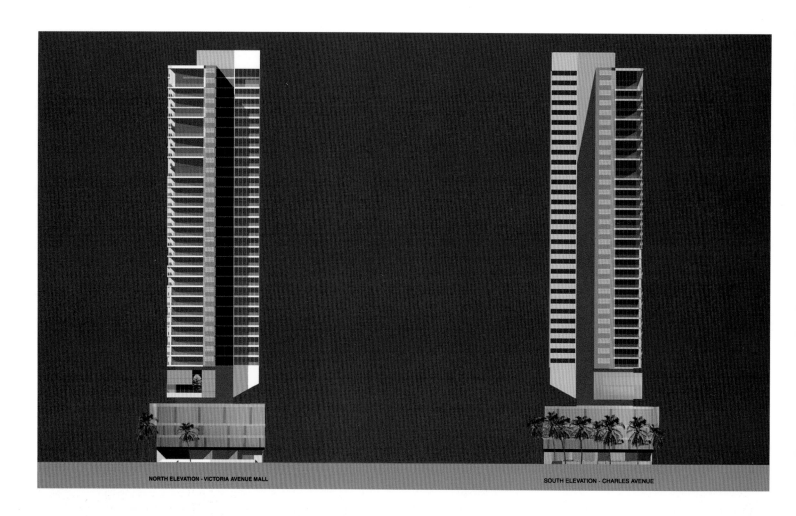

NORTH ELEVATION - VICTORIA AVENUE MALL

SOUTH ELEVATION - CHARLES AVENUE

规划布局及建筑设计

项目的主要建筑是由东面一座大型的菱形塔楼和西面的一座较小的矩形卫星楼组成。两座建筑通过电梯的升降机槽连接在一起，形成整个建筑的脊柱。建筑整体通过三个水平的和垂直的立面元素清晰而紧密地组合成一个整体。浓重的阴影线将不同部分在物理角度上区分出来。所有的立面各不相同，不同材质的运用和不同的表面处理反映出了不同的立面特点——向东开放，向西封闭，向北形成阴影。

建筑的设计还融合了被动式住宅的设计理念，使居住空间不仅全年舒适宜人同时还保持了较低的能耗。位于东面和北面的深嵌入式玻璃阳台最大程度上缓解了房屋的吸热情况，同时允许阳光在冬季可以直接照射起居室。西立面通过水平和垂直的百叶窗进行遮挡。建筑内所有的居住单元都保留了宽阔的开间高度，便于居室的自然采光。

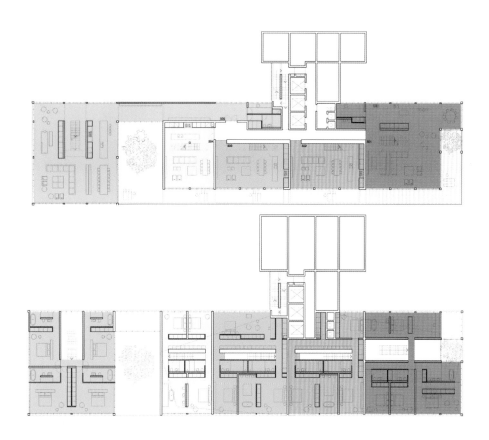

Profile

Located at the eastern extremity of the site, there are uninterrupted views of the beach and ocean to Coolangatta in the south and Surfers Paradise in the north. There are also significant views to the west of the Gold Coast hinterland and the Great Dividing Range.

Planning Layout & Architectural Design

The main building is composed of a large lozenge shaped tower to the east and a small rectangular satellite tower to the west. The two are linked by the lift core, which passes through the centre of the monorail turning circle providing the structural spine of the building off which the two towers cantilever. The building is clearly articulated by way of its three parts horizontal and vertical elements. Strong shadow lines emphasize the physical separation of the parts. All elevations are distinct but linked by a common kit of parts, with materials and finishes reflecting the different aspects of each elevation, open to the east, closed to the west, shaded to the north.

The design of the building has been developed to incorporate passive environmental strategies to create internal spaces with year round comfort for the occupants together with low energy consumption. To the north and east the deeply recessed balconies shade glazing to minimise heat gain, while allowing winter sun to penetrate living areas. Western elevations are shaded by the horizontal and vertical louvres. All units have large areas of floor to ceiling glazing allowing very high levels of natural daylighting.

FEATURE | 专题

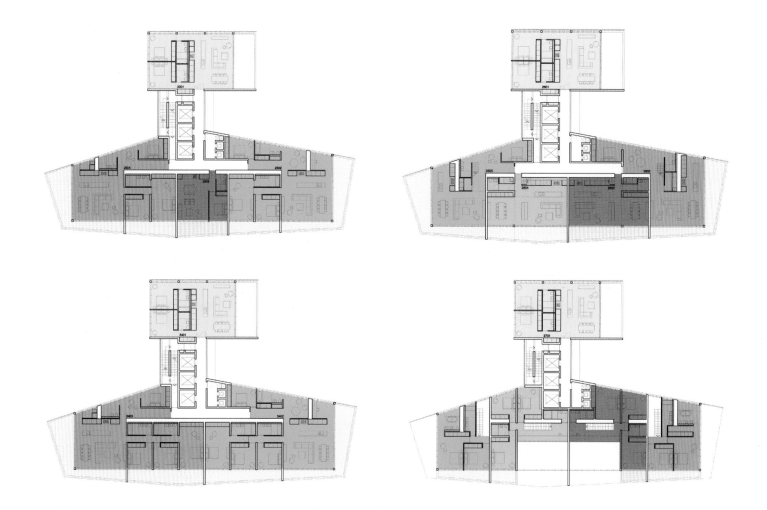

项目地点：波多黎各圣胡安
客　　户：互联公司
建筑设计：RTKL
项目负责人：汤姆·布林克
建筑面积：27 812 m²
摄 影 师：大卫·惠特科姆，卡洛斯·埃斯特瓦

PRACTICAL AND COMFORTABLE UPSCALE COMPLEX | Cosmopolitan

融实用性与舒适性于一体的国际化住区 —— 大都会

项目地点：波多黎各圣胡安
客　　户：互联公司
建筑设计：RTKL
项目负责人：汤姆·布林克
建筑面积：27 812 m²
摄 影 师：大卫·惠特科姆，卡洛斯·埃斯特瓦

Location: San Juan, Puerto Rico
Client: Interlink
Design Firm: RTKL Associates Inc.
VP-in-Charge: Tom Brink
Size: 27,812 m²
Photographer: RTKL/David Whitcomb, Carlos Esteva

项目概况

RTKL 根据互联集团的要求，设计了这个位于波多黎各圣胡安的高档住宅区项目。该项目让一度衰落的城区再次获得生机和活力。首期开发面积27 812 m²，共19层62个单位，包括标准间、联排别墅和顶层复式单位。

建筑设计

设计既融合了当地文化传统，又具有国际化色彩。整体采光好，通透性强。防飓风百叶窗玻璃不仅耐冲击性强，也兼具保证私密和安全性的特点。下电梯直通各户客厅，业主可在这里休闲娱乐或接待朋友。家具、厨房时尚现代，室外阳台更是家人聚会娱乐的好地方。工作服务人员设有专门的走道和电梯。

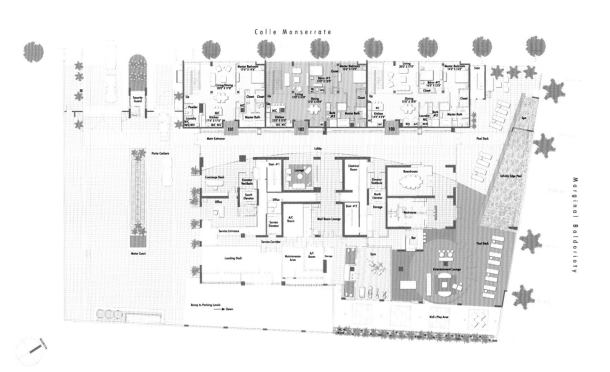

Profile

With RTKL's extensive portfolio in revitalization of dense, urban settings, the client Interlink Group was looking to create a new, upscale property in San Juan that would anchor and energize the once flourishing Miramar urban district. The Cosmopolitan reestablishes the standard of the Puerto Rican high-luxury market. The 27 812 m², 19-story tower features 62 standard, townhouse and penthouse units in the first phase.

Architectural Design

The RTKL design team nimbly embraces the local culture and tradition and enriches it with a global perspective. To capture natural light and openness, locally requisite continuous shear wall construction is replaced with a shear wall core and perimeter columns. Hurricane shutters are succeeded by high-impact resistant glass. To achieve exclusivity and security, a resident rides an individually keyed elevator from the garage to a private lobby in front of his door. Within each residence, guest entertainment and reception is discretely separated from family living and casual dining. Modern, furniture-inspired, Poggenpohl kitchens capture a sophisticated attitude. Balcony terraces extend outdoor living for intimate family use and more formal entertainment. The laundry and day quarters are accessed by domestic staff from the service corridor and separate elevator.

FEATURE | 专题

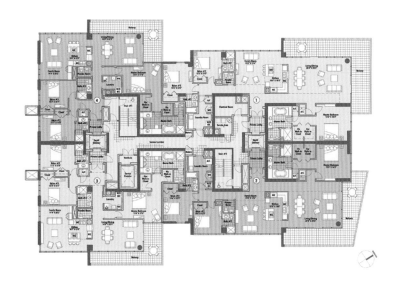

COMPOSED AND ELEGANT NORTH AMERICAN ARCHITECTURE

| Dalian Hungwei Lanshan, Period I

沉稳 优雅的北美风情建筑 —— 大连金州鸿玮澜山一期

项目地点：中国辽宁省大连市
建筑设计：筑博设计（集团）股份有限公司
用地面积：16 572 m²
建筑面积：84 959 m²

Location: Dalian, Liaoning, China
Architectural Design: Zhubo Design Group Co., Ltd.
Land Area: 16,572 m²
Floor Area: 84,959 m²

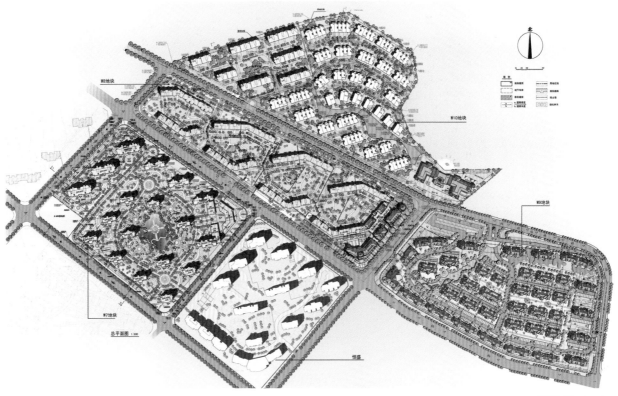

总平面图 Site Plan

NEW CHARACTERISTICS | 新特色

项目概况

项目用地位于辽宁省大连市金州区大黑山风景区西南侧。用地距离最近的干道是金州区的主要交通和商业干道五一路，南侧一路之隔紧靠大连理工大学城市学院，周边地块为尚待或正在规划建设中的住宅小区。用地交通优势明显。用地内有一条自然形成的冲沟穿越地块，为打造山水生活创造了天然条件。用地内几乎没有已建建筑，也给项目的设计提供了较大的创造空间。

规划布局

基于项目所在地块特定的城市区位以及紧邻大黑山自然风景区的优越自然条件，设计师们设想将山林、坡地等自然环境因素与当地原有的人文因素做有机的结合，提出以下设计理念：（1）充分利用城市发展契机，发挥场地优势条件，在尊重场地特点、保护原有生态环境的基础上，通过合理的规划设计，实现城市、

地下一层平面图　Plan for Basement Floor

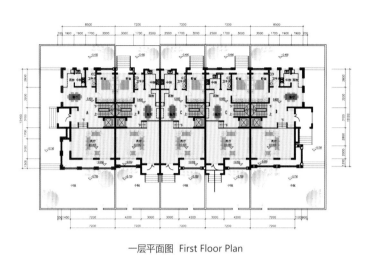

一层平面图　First Floor Plan

二层平面图　Second Floor Plan

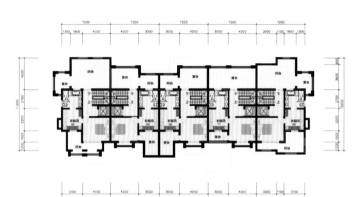

三层平面图　Third Floor Plan

NEW CHARACTERISTICS | 新特色

生态、居住结构的可持续发展。（2）以人的尺度作为设计基础，体现"以人为本"的人性化设计。

建筑设计

建筑立面设计为美式建筑风格。美式风格建筑天生适合山地地形，错落的地形有利于自然形成高低不同的体量和屋顶，与本基地自然坡地不谋而合。建筑主体色彩质朴雅致、比例和谐统一，体形错落有致、富于变化。采用变化丰富的暖色调，大量使用厚重的石材、朴实的原木、手工的瓦面等自然、野趣的材质。建筑细部运用线脚、窗套、铁艺栏杆，充分显示低密度产品的精工设计。利用面砖、涂料及金属、玻璃等材质的对比，体现建筑的有机性和丰富性。通过材质、颜色以及细部构造等多种设计方法的运用，使整个小区建筑风格丰富而统一，形成滨海城市里海滨住宅小区新颖独特的崭新建筑形象。

外观各不相同的一栋栋住宅，掩映在浓荫密林中，映射在碧水溪流上。在绿色的自然山体衬托下，形成了浓厚的山水风情氛围。

户型设计

一期项目主要以6层、7层花园洋房和9层、11层小高层住宅为主，并在一期的西侧规划了商业建筑作为配套设施。整体户型设计尊重北方生活习惯，户型设计中强调采光及保温的重要性。商业建筑设计强调其高效率的特点。

Profile

The project locates in the southwest of Da Heishan scenic spot, Jinzhou, Dalian, Liaoning. The nearest artery to the project is the main traffic road of Jinzhou and commercial artery of Wuyi Road. It adjoins the City College of Dalian University of Technology in the south; the neighboring area is yet to be or is being the residential area under construction. The advantage of transportation is obvious. The land has a natural gully through, which creates natural conditions for life with mountain and water. In the land, there are almost not existing buildings, also providing more space for the project design.

Planning and Layout

Based on the specific urban location and the next-door neighbor of Da Heishan scenic spot, the designer combines the natural environment of mountain, forest & sloping fields with the local cultural factors and puts forward the following design concepts: 1. Take advantage of the urban developing moment and bring the superior conditions of the place. Make reasonable planning and design to achieve the sustainable development of the city, ecological environment and residential buildings on the basis of keeping the local characteristics and the original ecological environment. 2. It should be people-oriented with humanized design.

Architectural design

The facade design applies the American architectural style. This kind of building is born to suit for mountain terrain that the random terrain can naturally form building blocks and roofs at different heights, happening to coincide with the base of natural slope. The main body of the building uses plain and elegant colors in a proportion of harmonious and unified, and its figure is well-proportioned and changeable. It also adopts rich warm colors, heavy stone, unadorned raw wood, handmade tile surface and other natural and wild materials. Architectural details use architrave, window cover, iron-wrought railing, fully displaying the low-density product design. The contrast of brick, coating, metal and glass materials embodies the organic attribute and richness of the building. The use of different kinds of design methods, such as material, color and detail structure and so on, makes the whole architectural style rich and unified, forming an novel and unique architecture image in the coastal residence in the seaside city.

Each residential building in different appearance sets in the jungle shade and reflects in the clear water stream. Against the background, it forms a strong landscape atmosphere.

House Layout Design

Project of Period I are mainly 6-floor or 7-floor garden houses and 9-floor or 11-floor high-rise residential buildings, keeping commercial buildings as supporting facilities in the west side. The whole house layout design adheres to the habits and customs of Northern China, focusing on the importance of light and heat preservation. Commercial building design emphasizes the characteristics of high efficiency.

MODERN, ELEGANT AND HIGH-QUALITY FRENCH-STYLE COMMUNITY

Greenland New Metropolis

现代简约的法式风格品质社区 —— 上海绿地新都会

项目地点：中国上海市崇明区
开 发 商：绿地集团
建筑设计：水石国际
项目规模：195 000 m²

Location: Chongming District, Shanghai, China
Developer: Greenland Group
Architectural Design: W&R Group
Size: 195,000 m²

项目概况

本案位于上海市崇明区，依托崇明岛"长江门户、东海瀛洲"的重要地位，同时尊享"田园水城，海岛花园"的优势资源，致力于打造富有生命力的现代化社区，创造高品质的宜居家园。项目选址北临乔松路，南靠翠竹路，与崇明区政府仅一街之隔，同时可受到新城市政配套规划的辐射，繁华的都市生活和便捷的商务体验，均可一站式全享。基地与东侧江帆路间为宽阔的城市绿化带，占据天然的景观资源。

规划布局

项目采用中轴对称的规则式布局，辅以与之相应的西式园林设计，彰显其庄重和尊贵感。一期北侧两地块以联排别墅和花园洋房为主，采用"组团半开放式"的基本空间结构。南北向两排洋房之间在满足日照、消防等基本规划条件的前提下，尽可能放大宅间距以打造稍带私密感与向心性的景观区域。由两侧的林间小径转而进入宅前屋后的集中式情境花园内，则是一番"别有洞天"的豁朗感；而联排别墅因间距较小则注重活泼的错落式排布，以穿插布置的点式绿化来打造"景缀宅间，宅隐景内"的场景，建筑与自然相互交融，浑然天成。

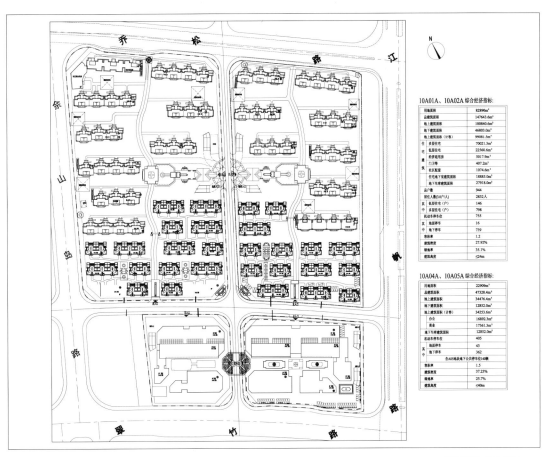

总平面图 Site Plan

NEW CHARACTERISTICS | 新特色

建筑设计

本案在设计初期便与材料商进行了充分沟通，期望通过低成本控制来打造雍容华贵的法式建筑风格。建筑除首层和门头采用石材外，其余建筑墙身皆采用真石漆，通过对分缝、分色等细节处的精心把控打造出高品质的立面效果。灰蓝色的孟莎式屋顶、淡雅的米色石材墙身、弧线形的老虎窗、精雕细琢的宝瓶栏杆，无不彰显法式建筑的浪漫、大气与尊贵。无论是多层洋房还是联排别墅，都严格遵循法式立面的构图比例，还原出柱式、老虎窗、山花等细节部品的轮廓与尺度。为了体现现代建筑的时尚感，设计中摒弃了古典法式建筑中繁缛的雕花及线脚，看似简约，但依然可以体味出浓浓的法式气息。

户型设计

本案中的住宅定位为法式风格，联排别墅的户型设计是带有创新性的亮点，设计师尝试打造"小面积，大享受"的经济型别墅空间。135 m² 两房两厅，宽敞有余，尊贵十足。大空间地下室，豪华入户玄关，全功能套房主卧，精心推敲空间尺度的舒适感，凸显别墅生活的高品质。而两层挑高客厅设计，三层出挑的大尺度景观露台，又为业主未来灵活改造提供了无限可能。每户独享的私家内庭院落，打造"家家有水，户户有花"的田园式生活场景。

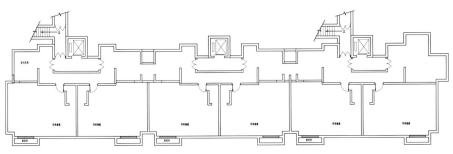

地下室平面图 1 Basement Floor Plan 1

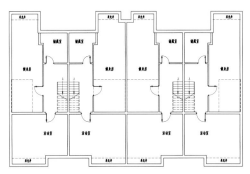

地下室平面图 2 Basement Floor Plan 2

NEW CHARACTERISTICS | 新特色

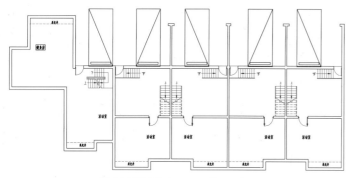

地下室平面图 3 Basement Floor Plan 3

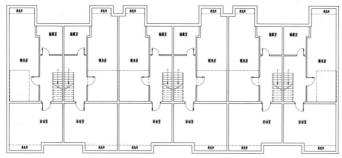

地下室平面图 4 Basement Floor Plan 4

Profile

The project is ideally located on Chongming Island of Shanghai City, which is known as "the gateway of Yangtze River and the east sea", enjoying beautiful landscapes and environment. It is envisioned to be an energetic modern community for high-quality life. The site is near to Qiaosong Road on the north, Cuizhu Road on the south and Chongming Government across a street, thus it benefits from the complete municipal facilities, providing one-stop services for both the daily life and business. Between the site and Jiangfan Road on the east, there is a wide green belt which provides great views for this project.

Planning and Layout

The buildings are arranged symmetrically with western-style landscapes in between, looking elegant and dignified. For the first phase, plots in the north are mainly for townhouses and low-rise garden houses which are arranged in semi-open style. Two rows of garden houses are designed with maximum spaces to meet the fire-fighting requirements and ensure enough daylight and privacy for the residents. The paths between landscapes will lead to the concentrated front yards or backyards, presenting another

world for people. For the townhouses, random layout is very important for small building distance. Green landscapes penetrate in between buildings, combining architecture with nature perfectly.

Architectural Design

In the early design stage, the architects had communicated with the material suppliers, hoping to create a luxury French architectural style with low cost. Except the first floors and the top of the doors, other walls are designed with stone coating. Jointing and color separation are carefully treated to create high-quality facades. Caeseous Mansart roofs, elegant beige stone walls, curved dormers, exquisite Aquarius rails……all demonstrate the romance, elegance and nobility of French-style architectures. No matter the multi-storey garden houses or the townhouses, all these buildings adopt French-style proportions in the facade design with columns, dormers, pediments and other details. On the other side, to highlight the modern sense, it abandons the complicated carvings and architraves in traditional French architecture, presenting the French style in an elegant manner.

House Layout Design

The houses here are designed in French style. The floor plans of the townhouses are innovative with "small area and great enjoyment". The 135 m^2 house type features two bedrooms, a sitting room and a dinning room, feeling dignified. Large-area basement, luxurious hallway, functional master bedroom and comfortable space proportion have highlighted the high-quality villa life. The two-storey high sitting room together with the protruding landscape terrace on the third floor provide limitless flexibility for the owner. Every house features a private courtyard which makes the dream of country life come true.

ECOLOGICAL AND ELEGANT NEOCLASSICAL URBAN COMMUNITY

| Liupanshui Future Cities

生态、典雅的新古典主义城市社区 —— 六盘水未来之城

项目地点：中国贵州省六盘水市
开 发 商：六盘水拓源集团康盛源德康房地产开发有限公司
规划设计：佰邦建筑设计公司（香港）
建筑设计：深圳市佰邦建筑设计顾问有限公司
占地面积：104 667.19 m²
建筑面积：377 176.05 m²

Location: Liupanshui, Guizhou, China
Developer: Liupanshui Tuoyuan Group Kangshengyuan Dekang Real Estate Development Co., Ltd.
Planning Design: P.B.A Architecture
Architectural Design: Shenzhen P.B.A Architecture
Land Area: 104,667.19 m²
Floor Area: 377,176.05 m²

项目概况

本项目用地位于贵州省六盘水市水城凤凰新区,地形为缓坡地,地势南高北低,高差约28 m。用地北侧为凤凰大道,西侧为水西南路,东面为水木清华高层住宅区;南面与玉源水厂隔墙相邻,西南角将有一条规划路——南二环路经过。

规划布局

规划中把绿地系统当成主要的公共活动空间渗透于居住生活的每一个角落,使人们在高强度、高节奏的工作之余得以更充分地投入社会交往,以及身心的调整。本方案结合原自然地貌,将地块分成南区和北区两个组团。北区景观以宅间前后院落为主,南区以环绕蜿蜒曲折的人工水系为主。结合周边自然的山景,南北区用一条步行道的景观主轴有机地联系在一起,步行道路设计成"中心绿化带",形成贯穿整个组团的绿化中心,将各个建筑之间连成一个有机的整体。

建筑设计

本方案在建筑形态上采用高层建筑沿街及南北向错落布置,共同构成整个小区,住宅建筑层数由16层到27层不等,将住宅环绕布置在小区的周边,内部空间最大化,形成内部小区共享环境最优化。建筑造型力求塑造全新的现代风格的城市社区形象,引入现代的材料和技术,采用明快的色彩,简洁流畅的造型,精致的细部处理,体现典雅厚重的新古典主义建筑风格;同时还进行了精细化的立面设计,在实际设计过程中结合当地的施工工艺情况、当地建材的实际情况等方面精心综合控制,绘制出指导施工的立面深化图,取得了不错的外观效果。

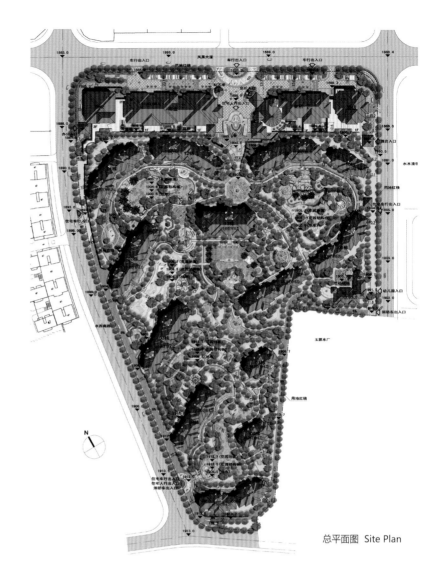

总平面图 Site Plan

NEW CHARACTERISTICS ｜ 新特色

户型设计

住宅户型多样，每户面积从80 m²至280 m²不等，从二房一厅至六房三厅不等。绝大部分户型朝南，基本都有景观，户内布局方正，功能分区清晰，景观阳台、服务阳台各在其位，客、饭厅分开。主卧房带套间卫生间，面积大的房型还设主卧衣帽间，大部分户型设有大面积花园式景观阳台，将花园绿意引入户内，也可在大阳台休闲小憩，欣赏城市山林和花园美景。

景观设计

在本小区中既营造了优美环境，又保持了空间的开敞和通透，使所有住户都拥有良好的朝向和景观。利用总体布局的空间结构，用三个台地标高解决场地高差，通过林荫步行道将整个小区景观资源紧密融合，有机联系创造点、线、面相结合的多层次绿化体系，加上生态水轴、景观铺地、蜿蜒的宅前小道、集中绿地、环境小品等一系列元素，使空间通透，绿荫葱葱，流水潺潺，一派休闲的优美景色。整个组团分为三个层次：中心绿化、组团绿化和宅间绿化，形成了点、线、面、体多样化丰富的立体绿色生态环境。

NEW CHARACTERISTICS | 新特色

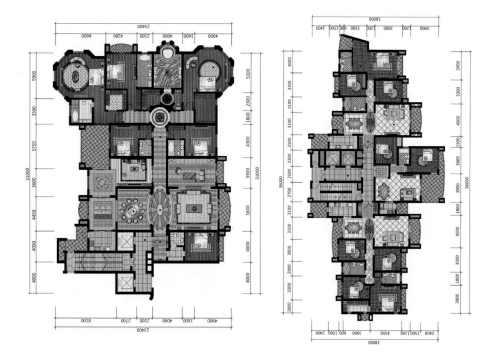

Profile

The project is located in the Phoenix New District, Liupanshui City, Guizhou Province, where the terrain is a gentle slope land, lower in the north and higher in the south with a height difference of 28m. It is south to Phoenix Road, east to Shuixi South Road, west to SMTH (the land of woods and waters) high-rise residential community and north to Yuyuan water plant. And there is a planning road, South 2nd Ring Road, passing by the southeast corner of the site.

Planning and Layout

Green space system planning as a major public space penetrates every corner of living life, so that people in spare time after high-intensity, high-paced work are being fully put into social interaction, as well as physical and mental adjustment. The program combines the original natural landscape, dividing the land into North and South, North District landscape

features the front and rear courtyard, South District features the surrounding artificial water system. Boasting the surrounding natural mountain views, both of districts connects the landscape axis with a footpath. As a central greenbelt, the footpath links up all the buildings as an organic integrity.

Architectural Design

In the form of architecture, high-rise buildings scattered along the street in north-south layout, together constitute the entire community. The residential floors are ranging from 16 to 27.

Residence is arranged in the periphery of the community to maximize interior space and optimize the interior environment. The architectural modeling seeks to shape the new modern image of urban communities, the introduction of modern materials and technology, the use of bright colors, simple and smooth styling, exquisite detail treatment, reflecting the heavy and elegant neo-classical architectural style; fine facade design combined with local construction techniques and the carefully integrated control of the actual situation of local building materials, creating a nice appearance.

House Layout Design

Houses vary from 80 m^2 to 280 m^2, from two bedrooms to six bedrooms and one living room to three living rooms in different collocations. Most of them are facing south with views, clear function division, separate kitchen and dining room, landscape balcony and serving balcony. Master bedroom has suite bathroom, and the large houses also have cloakroom in the master bedroom. Most of the units have large area for garden landscape balcony, on which you can enjoy the beauty of the mountains and gardens.

Landscape Design

It not only creates a beautiful environment, but also to keeps the space open and transparent, so that all households have good orientation and landscape. It takes advantage of the structure of the overall layout of the space, and solves the elevation height difference with three terraces, integrates the entire residential landscape resources through the tree-lined walkway closely, creates green system of multi-level in combination with points, lines and faces. In addition, there are ecological water axis, landscape paving, winding path in front the houses, concentrated green space and environmental embellishments. All these elements together picture nice sceneries. The whole residential group is divided into three levels: central green space, residential group green space and green space among houses.

NEW SPACE | 新空间

MODERN LOFT LIVING
Neptune Center

现代阁楼生活 —— 海洋中心

项目地点：中国四川省成都市
室内设计：香港优设计有限公司
主创设计：林湛远
设计团队：李明欣、蔡思杰、林卓敏
面　　积：60 m²

Location: Chengdu, Sichuan, China
Interior Design: YO DESIGN LIMITED
Chief Designer: Lin Zhanyuan
Design Team: Li Mingxin, Cai Sijie, Lin Zhuomin
Area: 60 m²

在实用面积60 m²的空间，打造出一个现代的简约风格设计，与淡淡的古典味，混然一体。

运用穿透与封闭并存的设计理念，穿梭于不同空间之中，彰显出整个设计的简洁明晰及宽敞、无边际的感觉。设计以木材为空间基调，局部配合精巧细致的镀银装饰元素，令空间内弥漫着现代与经典风格的味道，尤如阴阳的平衡，互相呼应。

多层次的视觉和触觉体验，营造温暖的氛围，流淌在一个严谨的设计环境中。

样板房户型图 Showflat Layout

NEW SPACE | 新空间

A net floor area of 60 square meters, the approach was to create a prestigious "Modern Loft Living" design with hint of classic touch in relationship with its bachelor-like aspects.

The design has been executed to manipulate this feel spacious. The materials as key components in a changing environment, the full-width sliding glass to the study gives an impression of extension of the living and dining room. Reflective materials such as mirrors and wall covering help in enlarging the space and making it glamorous. The lightness of timbers finishes also works as balancing of the Ying and Yang interpretation of the soul and warmth of materials brought forth through their own vital energy and complement each other.

Many-layered emotions can be experienced through sight and touch, creating an atmosphere of warmth flowing within a rigorous design setting.

DYNAMIC BUILDING WITH DIVERSIFIED FACADES | Basket Apartments in Paris

立面丰富多变 极具活力的建筑——巴黎篮子学生公寓

项目地点：法国巴黎19区
客　　户：Regie Immobiliere de la Ville de Paris
建筑设计：斯洛文尼亚OFIS建筑师事务所
项目团队：Team: Rok Oman, Spela Videcnik, Robert Janez, Janez Martincic, Andrej Gregoric, Janja del Linz, Louis Geiswiller, Hyunggyu Kim, Chaewan Shin, Jaehyun Kim, Erin Durno, Javier Carrera, Giuliana Fimmano, Jolien Maes, Lin Wei
面　　积：地块1 981 m²，小公寓35 m²，建筑931 m²，
　　　　　总建筑面积8 500 m²，景观1 050 m²
摄　　影：Tomaz Gregoric

Location: 19th district, Paris, France
Client: Regie Immobiliere de la Ville de Paris
Architects: OFIS
Project Team: Rok Oman, Spela Videcnik, Robert Janez, Janez Martincic, Andrej Gregoric, Janja del Linz, Louis Geiswiller, Hyunggyu Kim, Chaewan Shin, Jaehyun Kim, Erin Durno, Javier Carrera, Giuliana Fimmano, Jolien Maes, Lin Wei
Area: site 1,981 m²; size of studios 35 m²; building 931 m²; gross floor area 8,500 m²; landscape 1,050 m²
Photography: Tomaz Gregoric

项目概况

该项目位于巴黎19区的一块由Reichen & Robert事务所规划的狭长地带中，毗邻Vilette花园。新巴黎电车道打东北面经过，西南面靠近电车车库，往北是一块足球场。公寓墙壁的底下三层部分也是电车车库的墙壁。

项目的主要目的是为学生提供学习和交流的健康环境。沿足球场边是一个开放式课堂，同时可作为画廊使用，可以俯瞰花园或是远眺城市景观和埃菲尔铁塔。该画廊是去往公寓的必经之地。所有的小公寓尺寸相同，所采用的优化设计与施工的元素也相同，包括入口、浴室、衣柜、小厨房、工作台和一张床，而且每个小公寓都有一座阳台可以俯瞰街景。

建筑设计

10层楼高的狭长建筑很有存在感，根据功能需要每个体块都含有两个不同的立面，包含不用尺度室外阳台的立面用HPL木条做外皮，阳台体块随机变化，使外立面变得丰富、更具韵律感。变化的篮子外观创造了具有活力的外立面并消解了建筑的体量感。面向球场的立面有一条开放的走廊，从公寓入口通向立体金属网，楼的走廊连接两个建筑体块，是一个开放的公共空间。

可持续发展

该项目积极响应巴黎可持续发展的号召，计划入住后每平方米耗电50千瓦时或更少。项目施工进展顺利、节能性强、体量简单、隔热良好、一年四季通风自然、光照条件优越。外部走廊和玻璃楼梯有利于自然采光，既节省了能源又打造了舒适的社交空间。楼顶覆盖300 m²的太阳能光伏面板用来发电，收集的雨水用来灌溉户外绿地。

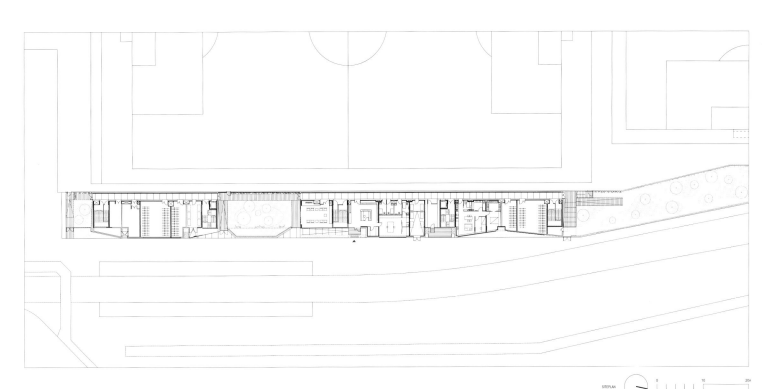

总平面图 Site Plan

NEW IDEA | 新创意

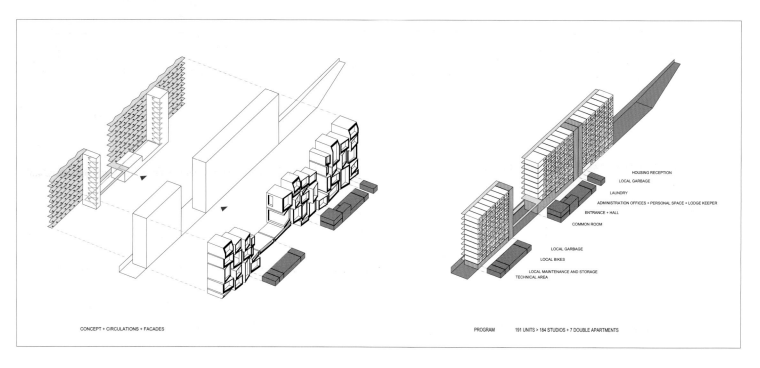

CONCEPT + CIRCULATIONS + FACADES

PROGRAM   191 UNITS > 184 STUDIOS + 7 DOUBLE APARTMENTS

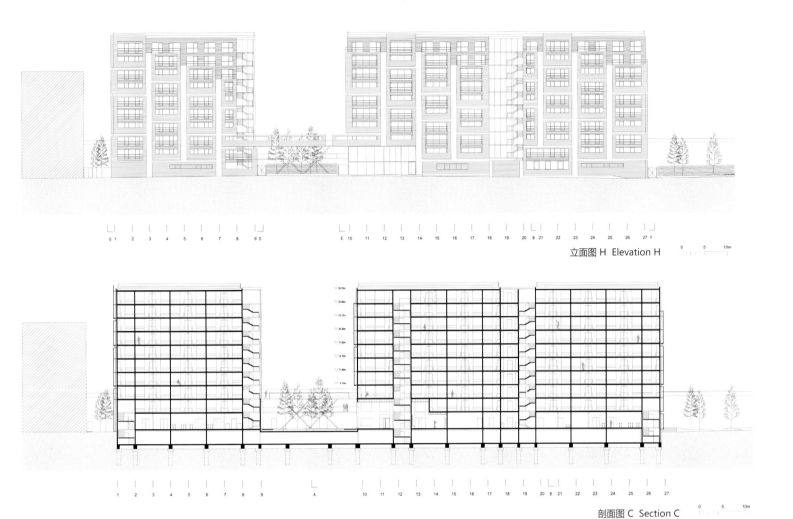

立面图 H Elevation H

剖面图 C Section C

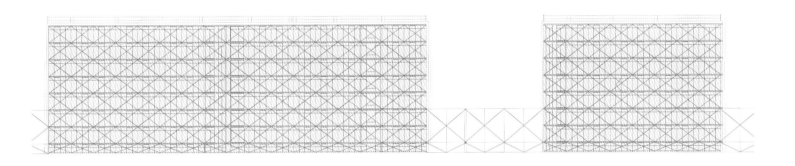

立面图 I Elevation I

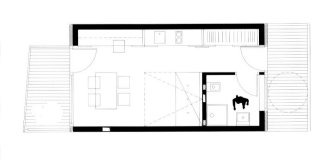

公寓平面图 Floor Plan of Apartment

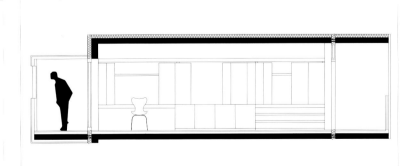

公寓剖面图 Section of Apartment

NEW HOUSE _111

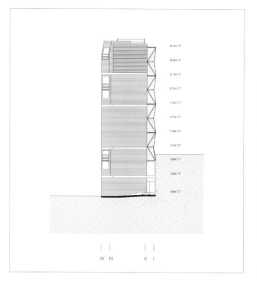
立面图 F Elevation F

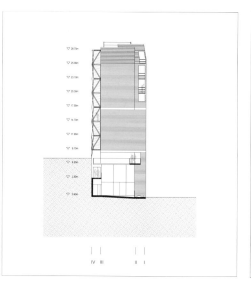
立面图 E Elevation E

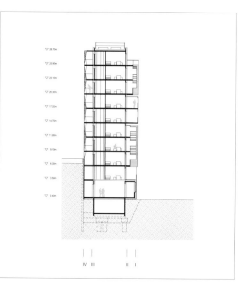
立面图 B Elevation B

NEW IDEA | 新创意

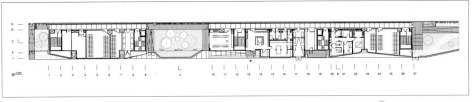
地面层平面图 Ground Floor Plan

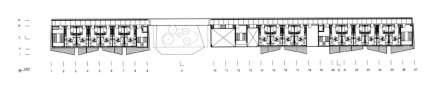
一层平面图 First Floor Plan

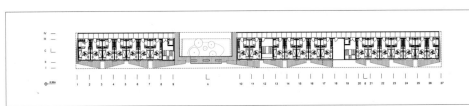
二层平面图 Second Floor Plan

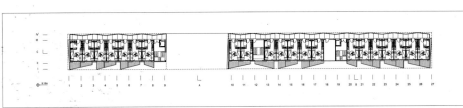
三层平面图 Third Floor Plan

Urban plan conditions

The project is located on a long and very narrow site, on the edge of Parc La Vilette in Paris's 19th district, within an urban development done by Reichen & Robert architects. On the northeast, new Paris tram route is passing along the site. The site is bordering with tram garage on the southwest, above which is a football field. The first 3 floors of the housing will inevitably share the wall with the tram garage.

The major objective of the project was to provide students with a healthy environment for studying, learning and meeting. Along the length of the football field is an open corridor and gallery that overlooks the field and creates a view to the city and the Eiffel tower. This gallery is an access to the apartments providing students with a common place. All the studios are the same size and contain the same elements to optimize design and construction: an entrance, bathroom, wardrobe, kitchenette, working space and a bed. Each apartment has a balcony overlooking the street.

Design concept

Narrow length of the plot with 10 floors gives to site a significant presence. Each volume contains two different faces according to the function and program: The elevation towards the street des Petits Ponts contains studio balconies–baskets of different sizes made from HPL timber stripes. They are randomly oriented to diversify the views and rhythm of the facade. Shifted baskets create a dynamic surface while also breaking down the scale and proportion of the building. The elevation towards the football field has an open passage walkway with studio entrances enclosed with a 3D metal mesh. Both volumes are connected on the first floor with a narrow bridge which is also an open common space for students.

Sustainable efficiency

The building is energy efficient to accommodate the desires of Paris' sustainable development efforts. The Plan Climates goal is that future housing will consume 50KW–h.m.² or less. The objectives of energy performance and the construction timetable were met by focusing on a simple, well insulated and ventilated object that functions at its best year round. Accommodations are cross ventilating and allow abundant day lighting throughout the apartment. External corridors and glass staircases also promote natural lighting in the common circulation, affording energy while also creating comfortable and well lit social spaces. The roof is covered with 300m² of photovoltaic panels to generate electricity. Rainwater is harvested on site in a basin pool used for watering outdoor green spaces.

DETAILED BRICK ARCHITECTURE WITH NUANCED SKIN | Scala, Amsterdam

立面丰富、表皮多变的精致砖结构建筑
—— 阿姆斯特丹斯卡拉大型住宅楼

项目地点：荷兰阿姆斯特丹
客　　户：Kristal/FarWest
建筑设计：荷兰FARO建筑师事务所
场地面积：6 900 m²
摄　　影：Luuk Kramer、Karel Tomei、FARO

Location: Amsterdam, the Netherlands
Client: Kristal/FarWest
Architectural Design: FARO architecten bv, bna
Site Area: 6,900 m²
Photography: Luuk Kramer, Karel Tomei, FARO

项目概况

在Bos & Lommer地区临近火车轨道的一片土地上，斯卡拉住宅楼胜利落成。项目基地所处位置极具挑战：两侧分别是铁路和繁忙的环城路。同时该项目也是Kolenkitbuurt区域重建的首个项目，具有重要意义。

重建工程旨在改善生活质量，在空间和建筑方面，加强其与A10高速公路另一侧的战前区之间的联系。如同铁轨沿线其他地方一样，"西方园林城市"的足迹开始踏上这块位于A10环线西侧的土地。

规划布局

FARO事务所为这个长230 m、宽30 m的地块设计了丰富的内容。项目包括六座6层或7层高的楼群及一座14层高的住宅塔楼，共提供208套各种类型的住宅单元。既包含私人购买的单位，又有出租单元，同时还有老年人生活辅助及关爱单元。规划的半地下停车场还将提供135个停车位。

靠近Leeuwendalerweg一端的是一座6层高楼。底层两个单元将设沿街出入口，使视线更加生动、开阔。建筑长度因退层结构产生细微变化，住户可以通过楼梯或电梯到达高层的公寓单元。

建筑设计

建筑灵感来自于上个世纪二十年代至四十年代精致的砖结构建筑。事实证明，橙色与褐色砖块加上白色点缀的战前建筑，一直以来都倍受青睐。通过FARO事务所的现代化诠释，这座建筑与阿姆斯特丹的城市精神完全契合，成为A10环路上一道引人注目的风景。

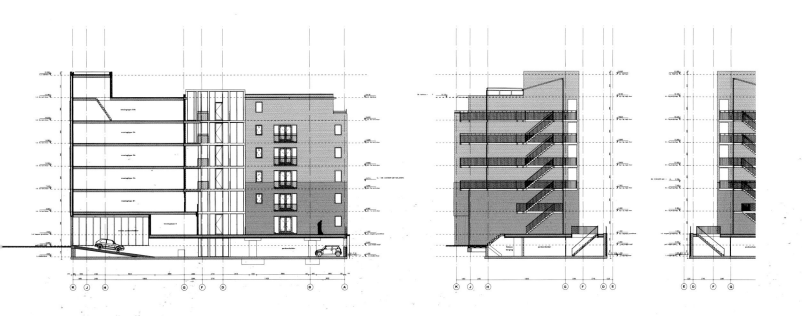

NEW IDEA | 新创意

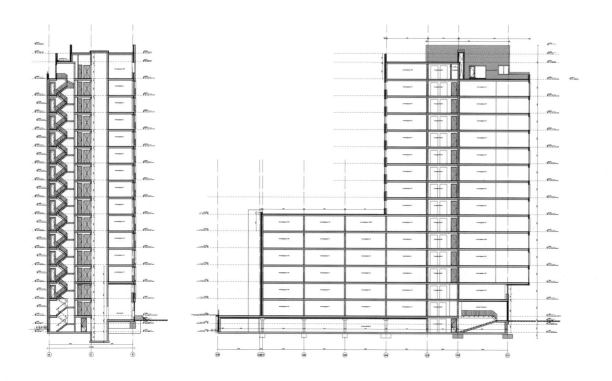

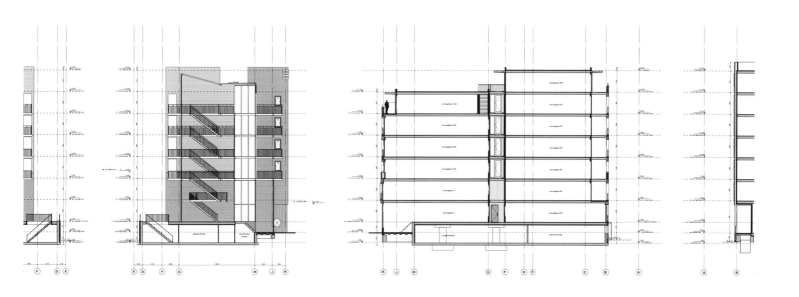

NEW IDEA | 新创意

Profile

On a strip of land next to the train tracks in the Kolenkitbuurt, Bos and Lommer, 'Scala' has been completed. It is a residential project sits on a challenging inner city site: between a rail way and a busy ring road. It is a major part of the restructuring of the pre-war Bos and Lommer.

Goal of the restructuring is to improve the quality of living and to improve the spatial and architectural connection with the pre-war part of Bos en Lommer on the other side of the A10. Like in so many places along the track, the world of the Western Garden cities now also starts on the west side of the A10 in the Kolenkitbuurt.

Planning Layout

FARO designed a very diverse program for the compact location of 230 by 30 meters. The project consists of 6 blocks of 6 or 7 layers and a residential tower of 14 layers. In total, there are 208 residences for varying occupants. There owner occupied units and rental units. Also assisted living and care units are included. The plan includes a semi-underground parking garage with 135 parking spaces.

On the Leeuwendalerweg, the block has 6 layers. The bottom two homes will have street access so the street view is more lively. The length of the block is nuanced by transparent recessions with stairs and elevators for the apartments on the higher layers.

Architectural Design

The architecture has been inspired by the carefully detailed brick architecture from the 20's - 40's. Pre-war architecture with orange and brown brick combined with white accents has proven to be appreciated over the years. According to FARO this Amsterdam idiom in its modern translation fits well in the city's ambition to jump over the A10 ring road.

COMME
BUILDINGS

商业地产

P126

阿姆斯特丹Mint酒店：
棱角突出的传统荷兰风格建筑

P134

泰州万达广场：
简洁现代、完整大气的新一代城市综合体

P140

FDA总部：
互动 高效 通达 简练

ON THE EDGE OF TRADITIONAL DUTCH STYLE ARCHITECTURE

| Mint Hotel Amsterdam

棱角突出的传统荷兰风格建筑 —— 阿姆斯特丹Mint酒店

项目地点：荷兰阿姆斯特丹东码头岛
客　　户：英国Mint酒店集团
建筑设计：英国Bennetts Associates Architects建筑事务所
建筑面积：28 000 m²
摄　　影：Peter Cook

Location: Eastern Dock Island, Amsterdam, the Netherlands
Client: Mint Hotel Group
Architects: Bennetts Associates Architects
Size: 28,000 m²
Photographer: Peter Cook

项目概况

　　阿姆斯特丹Mint酒店共有553个紧凑且高规格的卧室，还配备有餐馆、酒吧和会议厅。自从在威斯敏斯特成功引进一个屋顶休息室后，阿姆斯特丹也建起了空中休闲酒吧，大到可以容纳350人，可以俯瞰全城和阿姆斯特丹港。

规划布局

　　根据总体规划设计，酒店设置在一个庭院四周，卧室区在两层公共区上，公共区包含会议和餐饮设施。建筑的结构保证了卧室、露台和空中走廊都可以欣赏到城市大部分的壮观景色。坐在角落的大套房，可以欣赏到码头全景，中央庭院的设置有助于更多阳光进入卧室。公共区延伸到一个阳光充足的行人码头：北面，通往岛上的路镶嵌在建筑底边上，形成一个悬臂调节幅度，突出了建筑的棱角。

建筑设计

　　外立面由可移动的穿孔百叶窗、玻璃和砌砖构成，沿袭了传统的荷兰建筑精神。建筑每个部分材料的颜色都与周围不同的建筑风格相呼应。南向的立面所用的银蓝色砌砖彰显着本地的特色，而东西方向上的玻璃外立面和连接三个街区的桥梁反映了这个城市更现代的建筑语言。内部庭院外立面镀上了红色的锌，可以应对不同的光质。

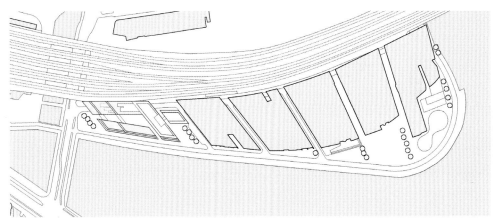

总平面图 Site Plan

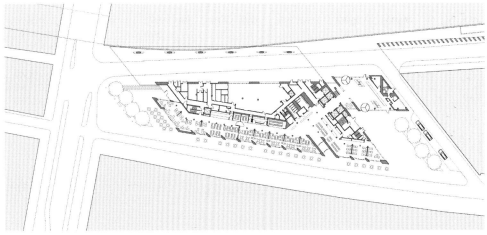

首层平面图 Ground Floor Plan

COMMERCIAL BUILDINGS | 商业地产

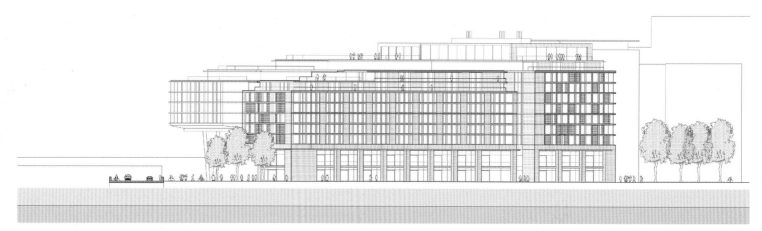

南立面图 South Elevation

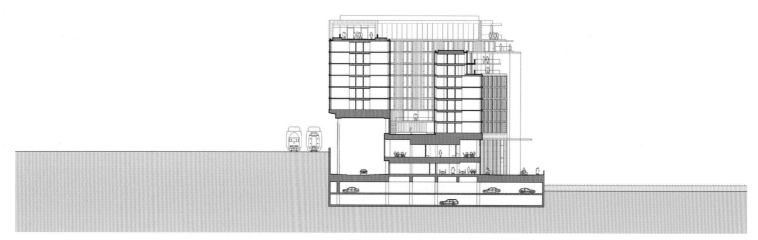

横切面图 Cross Section

Profile

Mint Hotel Amsterdam comprises 553 compact but high specification bedrooms, supported by a major restaurant, bars and conferencing. Following the successful introduction of a rooftop lounge at Westminster, the brief for Amsterdam included a SkyLounge bar large enough for 350 people, overlooking the old town and the port of Amsterdam.

Planning and Layout

In response to the masterplan, the hotel is arranged around a courtyard, with three linked blocks of bedroom accommodation rising above two public floors containing conference and dining facilities. The building's composition ensures that the bedrooms, terraces and SkyLounge make the most of the spectacular views across the water and the old town. The large bedroom suites, which sit in the angular corners of the building, enjoy panoramic views across the dock and the central courtyard allows high levels of daylight into the bedrooms. The public areas spill out onto a sunny pedestrian quayside facing the city: on the north side, the island's access road undercuts the building and produces a cantilevered range of accommodation that accentuates its angular plan.

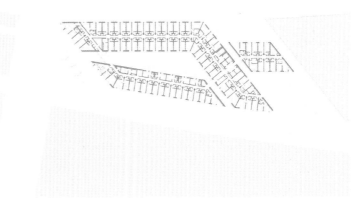

典型平面图 Typical Floor Plan

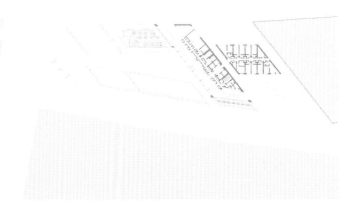

空中酒吧平面图 Sky Lounge Plan

COMMERCIAL BUILDINGS | 商业地产

COMMERCIAL BUILDINGS | 商业地产

Architectural Design

The facades consist of moveable perforated shutters, glazing and brickwork, emulating the spirit of traditional Dutch architecture. The palette of materials in each part of the building responds to the varying architectural styles which surround it. The south-facing elevation, with its silver blue brickwork responds to the vernacular of the old town it faces, while the glazing on the east and west facades and bridges between the three blocks reflects the architectural language of the more contemporary parts of the city. The internal courtyard facades are clad in red zinc, responding to the changing quality of light.

CONCISENESS AND MODERN, INTEGRITY AND ELEGANCE OF A NEW GENERATION OF URBAN COMPLEX | Taizhou Wanda Plaza

简洁现代、完整大气的新一代城市综合体 —— 泰州万达广场

项目地点：中国江苏省泰州市
开 发 商：万达集团
建筑设计：上海霍普建筑设计事务所有限公司
总用地面积：114 800 m²
总建筑面积：410 000 m²

Location: Taizhou, Jiangsu, China
Developer: Wanda Group
Architectural Design: Shanghai Hoop Architectural Design Co., Ltd.
Total Land Area: 114,800 m²
Total Floor Area: 410,000 m²

项目概况

泰州万达广场位于泰州市中心地段，东临海陵南路，西临青年南路，南临济川东路、北侧为通扬运河。规划总用地面积约为11.48万m²。总建筑面积约为41万m²，其中地上部分30.6万m²，地下部分10.58万m²。由购物中心、步行街、五星希尔顿酒店、写字楼、商务酒店、住宅、商铺组成的第三代城市综合体。

规划布局

在设计中从整体性原则考虑出发，考虑万达广场在泰州市地处要位，项目落成必将产生巨大的影响和效应，而本次设计中购物中心的建筑形象对于整个项目来说又是最重要的界面形象。因此在设计中通过整体统一的设计手法来达到完整的、大气的形象在周围风格迥异的建筑风格中突出自身的地标形象。

建筑及景观设计

泰州万达广场总体风格定位是简洁、现代、大气的建筑风格，设计风格大气磅礴，手法连贯，统一中有变化，富于时代气息，同时彰显尊贵典雅的气质。设计中突出了大商业立面的完整性和连续性，通过完整的体量和界面达到视觉的震撼，成为一个突出的界面来吸引人流。室外步行街在入口处采用与大商业呼应的设计手法，形成大体量和小体量之间的协调过渡。

立面图 1 Elevation 1

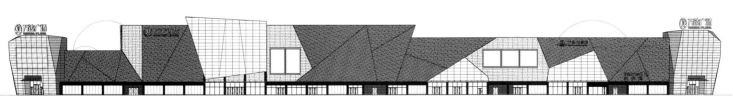

立面图 2 Elevation 2

COMMERCIAL BUILDINGS | 商业地产

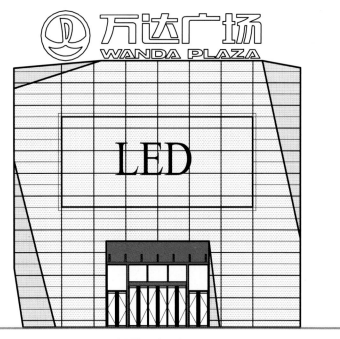

立面图2 Elevation 2

整个大商业部分进行了重点设计，以形成热烈，欢快，灵动，进取的气质，一方面满足商业氛围的需求，同时把握时代脉搏和万达企业文化的精髓。对于内部商业街主入口和各个二级入口都采用了重点突出的手法来强化处理，并且层级合理，另外在沿城市道路大商业界面与商业步行街的交叉口处重点设计，精细刻画，达到吸引人流进入步行街的作用。室外步行街采用标准单元段结合节点变化单元段的处理手法，在均质中追求变化。标准单元中窗的倾斜处理，产生了非常具有韵律感的视觉效果。与古建相邻的建筑采用花格窗的设计概念，局部设计了格栅状的金属装饰幕墙，与古建对话。

在设计中贯彻了"强调体块穿插，形成活跃明快的建筑形象"原则，在材料上以铝板和玻璃幕墙为主，从而确定了建筑大气雅致的基调，铝板采用氟碳彩涂铝板，玻璃幕墙亦在不同的部位采用不同颜色和反射率玻璃，通过材质的丰富变化形成活跃灵动的商业氛围，形成了体块感强、清晰明快、虚实搭配协调的整体立面风格。

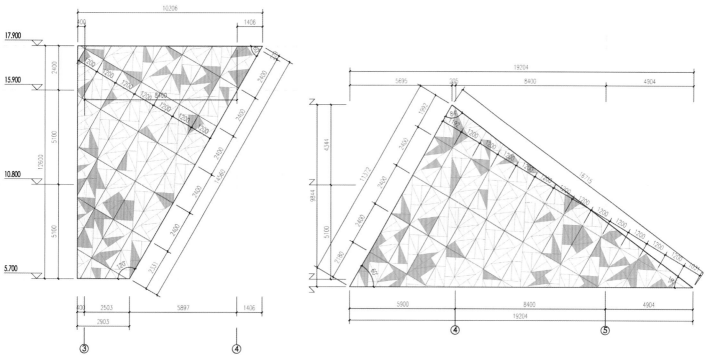

铝板幕墙立面大样图
Detail Drawing of Aluminum Curtain Wall Facade

COMMERCIAL BUILDINGS | 商业地产

Profile

Taizhou Wanda Plaza is located in the center of Taizhou, east to the South Hailing Road, west to South Youth Road, south to the East Jichuan Road and north to the Tongyang Canal. The project occupies a total land area of about 114,800 m^2, a total floor area of about 410,000 m^2, of which the ground part of 306,000 m^2, underground part of 105,800 m^2. The shopping center, Pedestrian Street, five-star Hilton hotels, office buildings, business hotels, residential area and shops comprise the third generation of urban complex.

Planning and Layout

The design of the project is based on the principle of integrity that the superior location of Wanda Plaza will produce great influence and effect, and the image design of the shopping center building is also the most important for the whole project interface image. Therefore, using the whole unified design technique to achieve the integrity and elegance will highlight the landmark image of the project around the buildings of different architectural style.

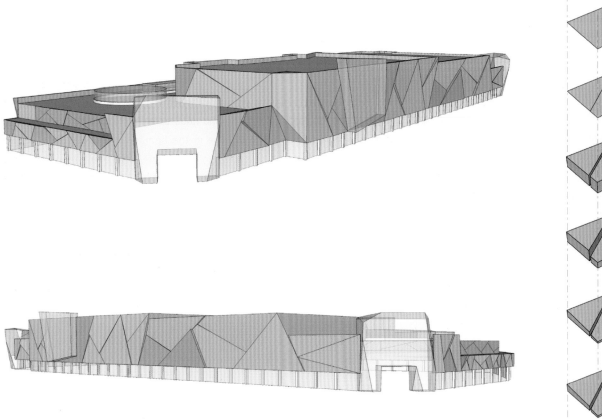

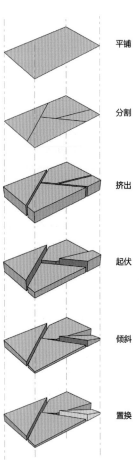

Architectural and Landscape Design

Taizhou Wanda Plaza is supposed to be concise, modern and elegant generally, and its design style is elegant with consistent technique, change out of unity, times spirit, and highlighting the noble and elegant temperament as well. The design highlights the integrity and continuity of the large commercial facades and achieves visual shock by the complete mass and interface to become a prominent interface to attract people. The outdoor Pedestrian Street at the entrance echoes well with the large commercial playground, forming a coordinating transition between the large scale and the small one.

The whole commercial part focuses on design in order to form a warm, lively, clever and aggressive temperament, which can not only create the business atmosphere, but also grasp the pulse of the time and Wanda's quintessence of enterprise culture. The internal commercial street entrance and each secondary entrance are strengthened to highlight with reasonable hierarchy. Besides, the intersection of the large commercial interface along the urban road and the commercial pedestrian street is designed subtly to attract people into the pedestrian street. Outdoor Pedestrian Street adopts both standard unit section and node changing unit section processing techniques, pursuing change in the homogeneous. Windows in the standard unit are tip-titled, thus producing the sense of rhythmical visual effect. And buildings adjacent to the ancient buildings apply lattice windows and grating-shaped metal curtain wall decoration partially to echo with the ancient buildings.

The design adheres to the principle of "emphasis on block penetration, form the active and lively architectural image", giving priority to the aluminum and glass curtain wall as materials, so as to determine the elegant and refined tone of the building. The aluminum sheet uses fluorocarbon painted aluminum plate and glass curtain wall also applies different color and reflectivity glass in different parts that create the active business atmosphere and form the strong feeling of body scale, clear and lively facade style.

INTERACTION HIGH PERFORMANCE MASTERY CONCISE | FDA Headquarters

互动 高效 通达 简练 —— FDA总部

项目地点：美国马里兰州
客　　户：美国食品和药物管理局
建筑设计：RTKL
面　　积：213 677 m²
摄 影 师：David Whitcomb，Paul Warchol

Location: White Oak, Maryland, USA
Client: General Services Administration
Design Studio: RTKL Associates Inc.
Size: 2300,000 SF or 213,677 SM
Photographer: RTKL/David Whitcomb and Paul Warchol

项目概况

目前,美国食品和药物管理局(FDA)管理着华盛顿特区18处40幢楼宇。作为1996年GSA卓越设计方案的一部分,RTKL与KlingStubbins合作共同设计FDA总部。FDA新总部选址于原海军水面作战中心,占地面积约526 091.33m²,总部共有9 000名工作人员。项目前期185 806.08m²工程已建成,整个项目预计于2014年完工。

规划布局

为了提高整体空间利用率和便于各部门之间互动,设计团队创造了一个总体规划,遵循大学校园建筑的公共庭院和行人空间的设计理念。总体规划是围绕中央的草坪上,促进FDA各部门交流的目的,同时也尊重个人隐私和工作的保密性,加强了建筑体跨学科合作的布局和设计。园区目前已建成66%,获得了多个LEED认证。

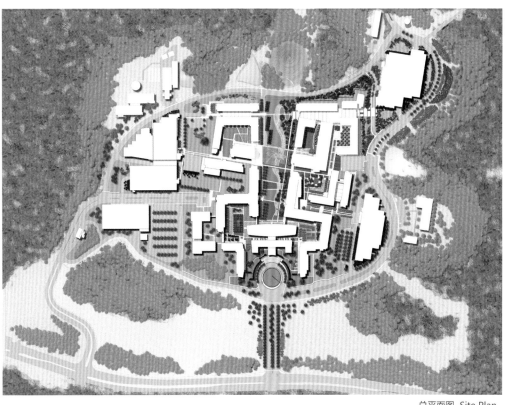

总平面图 Site Plan

COMMERCIAL BUILDINGS | 商业地产

Profile

Until recently, the U.S. Food and Drug Administration operated from 40 buildings in 18 locations throughout the Washington DC metropolitan area. In 1996 as part of GSA's Design Excellence program, RTKL was selected in collaboration with KlingStubbins to design the agency's new headquarters, consolidating its 9,000 employees on a 130-acre campus on the historic site of the Naval Surface Warfare Center. The resulting 2 million-SF development is being completed in phases, with final completion anticipated for 2014.

COMMERCIAL BUILDINGS | 商业地产

Planning and Layout

Challenged to improve overall space utilization and encourage more interaction among the FDA's core divisions, the design team created a master plan that follows the tenets of great university campuses, where buildings define a series of public courtyards and pedestrian spaces. The master plan is organized around a central lawn, with the placement and design of the buildings reinforcing FDA's mission of enhancing cross-disciplinary collaboration while also respecting the private and confidential nature of the organization's work. The campus is more than 66% complete and features several LEED certified buildings.

COMMERCIAL BUILDINGS | 商业地产

研发楼1 R&D Building 1

CLASSICAL, SOLEMN, COMPACT AND INNOVATIVE SPACE

Chengdu International Caizhi Sci-tech Park
古典、庄重、严谨的创意性空间——成都国际财智科技产业园区

项目地点：中国四川省成都市
开 发 商：成都睿敏置业有限公司
建筑设计：上海中建建筑设计院有限公司
总建筑面积：130 260 m²
地上建筑面积：99 620 m²
地下建筑面积：30 640 m²
容 积 率：2.49
绿 化 率：31.0%

Location: Chengdu, Sichuan, China
Developer: Chengdu Ruimin Properties Co., Ltd.
Architectural Design: Shanghai Zhongjian Architectural Design Institute Co., Ltd.
Total Floor Area: 130,260 m²
Floor Area Over Ground: 99,620 m²
Floor Area Under Ground: 30,640 m²
Plot Ratio: 2.49
Greening Ratio: 31.0%

项目概况

本项目位于成都市东三环外，龙泉驿区成龙路以南、博美银河装饰城以东，临近四川师范大学龙泉校区，是成都市区和龙泉驿区的交界处。建设基地面积为4万m²，拟建的建筑为两幢地下二层、地面38层的科研楼、配套附属建筑及28幢创意性研发楼及生产楼。

规划布局

在基地北部设置38层超高层科研楼，主要面向龙泉驿区正在迅速扩展的汽车产业、装饰产业以及越来越多的公司总部的迁入，为提升本区的经济和社会整体品质奠定良好的基础。考虑到近期周边的企业规模，标准层可以灵活划分，并以每单元不少于300 m²的三单元格局为主体。

南部设置低层创意性研发楼，每栋楼建筑面积300 m²~500 m²，面向目前蓬勃发展的低污染、高品质、高效率的创意性产业，为经济转型提供良好的产业基地。中部为低层创意性生产楼，根据规划要求用地比例不小于总用地的20%，与南部的创意性研发楼配套使用。

建筑设计

地下部分：结合周边道路将地下车库的出入口布置于基地北侧科研楼的东西两端，而将主要的设备及辅助用房合理布置于地下一、二层，在满足安全疏散的前提下，合理布置车位和行车路线，确保较为宽松的车行宽度、进退距离和转弯半径满足行车和停车要求；同时也使地下一层有规整的平面，满足其它服务功能。

地面部分：超高层科研楼主要由三层裙房及2栋超高层组成，3层裙房主要以科研配套附属用房功能为主，既能满足主楼的日常需求，又能服务整个基地，良好的辐射整个小区。在满足消防和疏散的前提下，均匀布置垂直交通系统，将大面积的规整空间预留给商业服务。在科研配套附属用房靠中央核心区一至三层设置一个垂直式中庭，既优化了平面与垂直交通的组织，又增加了空间的恢弘大度和使用价值。充分利用裙房屋顶空间，将其改造成有利于与中央景观轴衔接的屋顶观景平台，同时也有效地改善了上部科研的综合环境。4~38层平面设计充分考虑了科研用房的功能要求，使各层均可灵活组合，同时兼顾高层建筑的消防和安全疏散要求。避难层分别设置于18及33层屋顶。

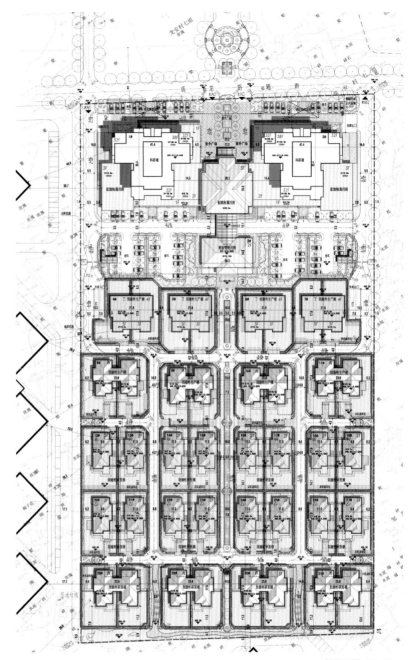

总平面图 Site Plan

总体鸟瞰图 Site Plan (bird-view)

COMMERCIAL BUILDINGS | 商业地产

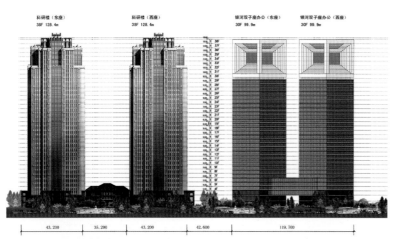

沿成龙路建筑立面分析图　Facade Analysis along Chenglong Road

科研楼立面图　Elevation of Science Research Building

Profile

Located outside the east 3rd ring road of Chengdu City, on the south of Chenglong Road of Longquanyi District and on the east of Bomei Galaxy Decoration City, it is near to Longquan campus of Sichuan Normal University at the urban-rural fringe. Occupying a land area of 40,000 m², there will be two scientific research buildings(two floors under ground and 38 floors on ground), supporting facilities, and 28 innovative R&D and manufacturing buildings.

Planning and Layout

The 38-storey scientific research block is set in the north of the site, mainly for the rapidly developing auto industry, decoration industry and increasing head offices. It will promote the economic and social development of this area. In consideration of the corporation sizes of the surroundings, the standard floor can be divided flexibly. Three-unit layout is normal and each unit will be no smaller than 300 m².

The lower 300-500 m² R&D buildings are built in the south for the innovative industries of low pollution, high quality and high efficiency. They will be a good platform for economic transition. And the low-storey manufacturing buildings are in the center, occupying 20% of the site according to the planning requirement. They will serve together with the R&D buildings in the south.

研发楼2　R&D Building 2

研发楼3 R&D Building 3

Architectural Design

Underground: Based on the surrounding traffic system, the exits and entrances of the underground parking are set in the east and west end of the scientific research blocks. The equipment rooms and other rooms are set on the basement 1 and 2 floor. It not only ensures safe evacuation but also well arranges the parking lots and drive ways to meet different requirements. At the same time, it provides a flat basement 1 floor for other functions.

On Ground: The scientific research block is composed of a three-storey podium and two high-rise buildings, of which the podium is served as the supporting facility for scientific research of the block itself and the whole sci-tech park as well. Vertical traffic system is set with enough spaces for firefighting and evacuation. There is a three-floor vertical patio in center of the podium which well organizes the traffic system and maximizes the space values. It also takes advantage of the roof space of the podium to transfer it to a landscape platform which will connect with the central landscape axis. The landscape platform will also improve the environment of the upper floors. The floor plan for the 4th ~38th floor is flexible to meet different scientific research requirements, and at the same time reaches the standard for firefighting and safe evacuation. The refuge floors are set on the top of the 18th and 33rd floor.

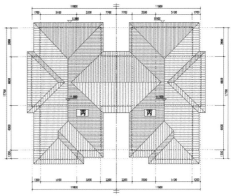

屋顶层平面图 Plan for Roof Floor

1-1 剖面图 1-1 Section

NEW ERA 北京新纪元建筑工程设计有限公司

云冈石窟博物馆

地址：北京市海淀区北小马厂6号华天大厦20层
电话：010-63483388
　　　58891220
传真：010-63265003
邮箱：xinjiyuanscb@126.com
网址：www.bjxinjiyuan.com

建筑·新纪元　设计·新生活

设计师广场　　　中央美术学院美术馆

建外SOHO

司简介

京新纪元建筑工程设计有限公司是建设部颁发具备"建筑工程设计甲级"股份制企业。公司有近500余人的设计团队，有多名国内知名的行业一级注册建筑师、一级注册结构工师，以及由多名国内著名的设计大师组成的专家顾问队伍。筑师特点鲜明、综合能力强、经验丰富、获奖作品众多。在共建筑、居住建筑、小区规划、场馆建筑、商业建筑、室装饰、园林景观等方面具有特色和较强竞争力。

公司重组以来注重吸取、借鉴国外建筑设计的先进理念和做法，与国际著名设计大师合作完成了多项设计作品并获得优秀设计奖项。

如：建外SOHO（与日本合作）、天鸿幸福广场（与意大利合作）、设计师广场（与日本SAKO设计工社合作）、中央美术学院美术馆（与日本矶崎新事务所合作）、石家庄西美-第五大道（与加拿大诺杰国际建筑师事务所合作）、唐山景泰翰林（与德国合作）等项目。

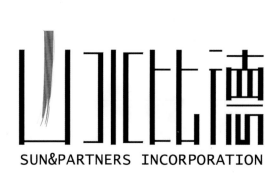

PROFESSION CHANGE LIFE

广州山水比德景观设计有限公司

湘江一号实景图

广州市天河区珠江新城临江大道红专厂F19栋
TELL:020-37039822 37039823 37039825
FAX:020-37039770
E-MAIL:SSBD-S@163.COM

服务对象：

广州珠江新城·珠江北岸 文化创意码头

领航创意 共创未来

广州市天河区珠江新城临江大道685号红专厂F19栋
TELL:020-37039822 37039823 37039825
FAX:020-37039770
E-MAIL:SSBD-S@163.COM

全国招募

TEL:020-37039822
FAX:020-37039770